LF

THE PHYSICS OF MUSIC

ALEXANDER WOOD'S

The Physics of Music

revised by
J. M. BOWSHER
Lecturer in Physics, University of Surrey

SEVENTH EDITION

LONDON

CHAPMAN AND HALL

A Halsted Press Book
John Wiley & Sons, Inc., New York

First published 1944
by Methuen & Co. Ltd.

Fifth edition 1950
Sixth edition 1962
Seventh edition 1975

Reprinted 1976
published by Chapman and Hall Ltd
11 New Fetter Lane, London EC4P 4EE

© *1975 Chapman and Hall Ltd*

Printed in Great Britain at the
University Printing House, Cambridge

ISBN 0 412 13250 8 (cased edition)
ISBN 0 412 21140 8 (Science Paperback edition)

Distributed in the U.S.A. by Halsted Press,
a Division of John Wiley & Sons, Inc., New York

Library of Congress Cataloging in Publication Data

Wood, Alexander, 1879–1950.
  The physics of music.
  Bibliography: p.
  1. Music—Acoustics and physics. I. Bowsher, J. M. II. Title.
ML3805.W665   1975   781'.1   74-4025
ISBN 0-470-95978-9 HB;   0-470-15057-2 PB.

PREFACE TO FIRST EDITION

I HOPE that this little book may serve as an introduction for some to the very interesting borderland between physics and music. It is a borderland in which the co-operation of musicians and physicists may have important results for the future of music.

The typescript and proofs have been read by Miss Nancy Browne from the point of view of the general reader, and many obscure passages have been clarified. On the technical side I am indebted to Dr Pringle, who has read the proofs and given me valuable criticism and advice. Miss Cawkewell has helped me with the illustrations, Mr Cottingham has supplied the photographs for Figs. 1.7 to 1.10, and my secretary, Miss Sindall, has been responsible for the typing and for the assembly and preparation of the material. Because of the help received from these and others the book is a much better book than it would otherwise have been. For its remaining imperfections I must take full responsibility.

I should like also gratefully to acknowledge the patience and consideration of the publishers and the way in which they have grappled with the difficulties of the production of a book of this kind in war-time.

ALEX WOOD

PREFACE TO THE SEVENTH EDITION

SINCE the sixth edition, Dr Wood's book has remained a firm favourite both amongst my own students at the University of Surrey and those at other universities and colleges. Despite the many excellent new books published since 1961, *The Physics of Music* retains its place because it seems to have just the right balance to convey the maximum amount of useful information to a student with the minimum of displeasure.

I have, therefore, contented myself in this new edition with changing the notation of pitch intervals from the savarts used by Wood (which have caused much confusion) to the more widely used cents, and at the same time, recalculating all the values given

in the text to remove a few errors which have come to light.   I
have also, as far as practicable, changed over to the metric system
of measurement in view of the fact that the old feet and inches are
no longer being taught in our schools, and made some very minor
alterations in the section on recording and reproduction of sound.

J. M. BOWSHER

Guildford
June, 1975

CONTENTS

ILLUSTRATIONS

ACKNOWLEDGMENTS

Thanks are due to the authors, publishers, and manufacturers named below for permission to reproduce the following illustrations:—

Figs. 1.1, 1.11, 2.4, 3.3, 5.1, 5.2, 5.3, 5.4, 5.5, 5.8, 8.12, 12.1, 12.2, and 12.5 are from *Sound Waves and Their Uses* by A. Wood (Blackie & Son, Ltd.).

Figs. 1.2 and 1.13 are from *A Complete Physics for London Medical Students* by W. H. White (R. Clay & Sons, Ltd.).

Figs. 1.6, 8.1, 8.5, 12.4, 12.6 *a* and *b*, 12.8, 13.2, and 13.3 are from *Acoustics* by A. Wood (Blackie & Son, Ltd.).

Fig 2.3 is from *Science of Musical Sounds* by D. C. Miller (Macmillan, N.Y.).

Figs. 5.6 and 5.7 are from *Sound Waves, Their Shape and Speed* by D. C. Miller (Macmillan, N.Y.).

Figs. 3.1 and 3.2 are from *Applied Acoustics* by Olson and Massa (The Blakiston Co.).

Figs. 4.1, 7.2, 7.3, 8.13, 8.14, 9.2, 9.3, 9.9, and 11.3 are from *The Journal of the Acoustical Society of America*, 1931, 1934, 1937, 1938, 1939, and 1942.

Figs. 4.2 and 9.4 are from *The Psychology of Music* by Seashore (McGraw-Hill Pub. Co.).

Figs. 5.9, 6.1, 6.2, and 6.4 are from *Speech and Hearing* by Fletcher (D. Van Nostrand Co., Inc.).

Fig. 6.3 is from *The Mechanism of the Cochlea* by Wilkinson and Gray (Macmillan).

Fig. 6.5 is from *Hearing : Its Psychology and Physiology* by Stevens and Davis (John Wiley & Sons, Inc.).

Fig. 7.1 is from *Physical Basis of Piano Touch* by Ortmann (Kegan Paul).

Figs. 7.4 and 10.1 are from *Sound* by A. T. Jones (D. Van Nostrand Co., Inc.).

Fig. 7.5 is from *Sensations of Tone* by Helmholtz (Longmans).

Figs. 7.6 and 7.9 are from the *Proceedings of the Indian Association for the Cultivation of Science.* 1920–1.

Fig. 7.7 is from *The Philosophical Magazine 1916* (Taylor & Francis).

Figs. 7.8, 7.10, 7.11, and 7.12 are from *A Scientific Search for the Secret of the Stradivarius* by Saunders. (The Franklin Institute).

Fig. 8.2 is from *Acoustics of Orchestral Instruments* by Richardson (Edward Arnold & Co.).

Fig. 8.3 is from *A Textbook of Sound* by A. B. Wood (George Bell & Sons, Ltd.).

Fig. 8.4 is from *Science and Music* by Jeans (C.U.P.) from a paper by G. J. Richards.

Figs. 8.6 and 8.7 are from the *Proceedings of the Physical Society* (University College, London).

Figs. 8.8 and 8.9 are from *Modern Organ Building* by Lewis (Reeves).

Figs. 8.10 and 8.11 are from the *Proceedings of the University of Durham Philosophical Society*, 1938.

Fig. 9.1 is from *Human Speech* by Paget (Kegan Paul).

Figs. 9.5, 9.6, 9.7, 9.8, 9.10, 9.11, 9.12, and 9.13 are from *Musical Wind Instruments* by Carse (Macmillan).

Fig. 9.14 is from *Traité d'Acoustique* by Chladni.

Fig. 9.15 is from *Carillon Music and Singing Towers of the Old World and the New* by W. G. Rice (Dodd, Mead & Co.).

Fig. 9.16 is from *Church Bells* by H. B. Walters (O.U.P.).

Fig. 11.1 is from a paper in the *Philosophical Magazine* by A. F. Dufton (Taylor & Francis).

Fig. 12.3 is by Science Museum.

Fig. 12.7 is from *Electronics*, May, 1940 (McGraw-Hill Pub. Co.).

Fig. 12.9 is by Acoustico Enterprises Ltd.

Fig. 12.10 is by F.W.O. Bauch Limited.

Fig. 12.11 is by Acoustical Manufacturing Co.

Fig. 12.12 is by Metrosound.

Figs. 13.1, 13.4, 13.6, 13.7, 13.8, 13.10, 13.11, 13.12, and 13.13 are from *Planning for Good Acoustics* by Bagenal and Wood (Methuen & Co., Ltd.).

Fig. 13.14 is from *Architectural Acoustics* by Knudsen (John Wiley & Sons, Inc.).

CHAPTER 1

THE NATURE OF SOUND

THE TERM sound is sometimes applied to the quite definite and specific sensation which we associate with the stimulation of the mechanism of the ear. It is also applied to the external cause of the sensation. In this book, we shall be concerned mainly with the latter meaning of the term, and the context will usually avoid any confusing ambiguity. But while our attention will be directed to what is happening outside the ear, it is the sensation produced by the external causes which is our sole standard of judgement. It is the character of the sensation which determines whether a sound is *musical*, and therefore comes within our present scope, or unmusical, and therefore lies outside it.

Musical sounds are those which are smooth, regular, pleasant, and of definite pitch. Unmusical sounds are rough, irregular, unpleasant, and of no definite pitch. The classification is only approximate at the best. Many sounds which pass as noises have associated musical notes, and almost all musical notes have associated noises which we deliberately ignore in order that we may concentrate our attention on the note. In spite of this, however, the classification remains a useful one, and it is to the musical sounds in this sense that we shall confine ourselves in this book.

The source of a musical note is always some ' system ' in vibration. It may be a stretched string, as in the piano or violin, a column of air, as in the organ pipe or a wind instrument, a bent rod, as in a tuning-fork, or a straight bar, as in a xylophone. These vibrations communicate themselves to the air in contact with the vibrating system, and this to-and-fro vibration is communicated through the air from the source to the ear of the observer. We have thus three principal aspects of the musical note to consider: (*a*) the vibration of the source, (*b*) the transmission through the medium, and (*c*) the reception by the hearer. Our inquiry will be mainly physical, but we shall find it impossible to avoid crossing the border occasionally into territory which properly belongs to anatomy, physiology, psychology, or æsthetics.

Transmission of Sound through a Medium.—The transmission of sound-waves can be made much more intelligible if we compare it with other similar phenomena which can be followed by eye. At first sight there seems to be no similarity at all between the propagation of waves on the surface of water and the transmission of sound through the atmosphere; but the similarity is very close indeed, and, as water-waves and ripples are easily observed, their behaviour helps us a good deal in the study of sound-waves. Now, the essential thing about a water-wave is that a ' condition ' (motion, elevation, depression, &c.) is transmitted from point to point by a 'medium' (the water surface) without movement of the medium as a whole in the direction of the waves. The illusion that the water is travelling with the waves may be very strong as we watch them, but every sea-swimmer knows that while the water at the crest is travelling with the waves, the water at the trough is travelling against the waves and there is no forward movement of the water as a whole. Now, in the same kind of way sounds are carried from point to point through the air without movement of the air as a whole in the direction of propagation of the sound. One of the first experiments tried by Guericke (1602–1686) after he invented the air-pump was that of causing a bell to strike in a vessel from which the air was being exhausted. He showed that the sound became fainter and fainter as the vacuum became more perfect. The experiment was repeated by Robert Boyle (1626–1691), who concluded that ' whether or no the air be the only, it is at least the principal medium of sounds '. Guericke found it difficult to understand how sound could be carried by air which is at rest. It is, of course, only apparently at rest. It is not moving as a whole in the direction in which the sound is being carried—it is moving to and fro. Guericke's difficulty disappears when we watch water-ripples starting from some object moving on the surface of the water and spreading across the surface, which itself moves up and down, while the wave passes onward.

The relation of the motion of the wave to the motion of the medium is made clear by a wave-model such as that shown in Fig. 1.1, Plate I. The balls move up and down while the wave moves forward, any given ball being at the top of its path as the crest passes and at the bottom as the trough passes. This model also illustrates the definition of some important features of wave-motion. The maximum displacement of the medium, during the passage of a wave, from its position of rest is the *amplitude* of the motion. It is the height of the crest or the depth of the trough measured from the mean position. The time taken by any point to complete one vibration is the *period* of the motion, and the number of vibrations completed in one second is the *frequency*. Thus if the period is T and the frequency f, then $T = \frac{1}{f}$. The stage which a particle has reached in its vibration is its *phase*. Thus a crest and a trough are said to be in opposite phase or two crests to be in the same phase. The *wave-length* λ is the distance between two successive particles in the same phase—i.e., two crests or two troughs. The *particle velocity* is, as the name suggests, the velocity of a particle of the medium as distinguished from the wave-velocity which is the velocity of the wave form. The word particle must not be interpreted too literally. In the case of the model its significance is clear, but in the case of sound-waves the molecules are much too small to act as particles in the sense in which we are using the term. In the case of air-waves we must think of masses of air very tiny, it is true, but containing millions of molecules.

A very important relation, true for all kinds of wave-motion, can be derived from consideration of the model or of the wave-motion it represents. Any particle oscillates up and down f times per second. Every time it reaches the top of its path the crest of a wave passes it. Therefore f waves pass it in one second. The first of these will, by the end of the second, be f wave-lengths away. It has therefore travelled a distance $f \lambda$ in one second, and this is by definition the velocity of the waves, so that $c = f \lambda$ where c is the velocity of the waves.

Looking at the model when it is working, we notice that each ball describes exactly the same vibration in exactly the same time, but that each ball is slightly ahead of its neighbour farther from the source of the waves and slightly behind its neighbour nearer the source. This is what is meant by saying that when waves are being propagated the phase changes continuously from point to point along the line of propagation.

The waves we have considered so far have been 'transverse'. In the water-waves the surface has moved vertically up and down while the wave has been propagated horizontally along the surface. In the case of the model also, the balls execute a vertical motion the wave travelling horizontally. In both cases the motion of the 'particles' is at right angles to that of the wave. But in the case of sound-waves the matter is different. The waves are 'longitudinal' in type, which means that the motion of the particles is to and fro in the line of propagation. Thus the motion

FIG. 1.2.—Wave diagram to illustrate propagation of waves

of the particles is *parallel* to the motion of the waves. The motion can perhaps best be understood by using Fig. 1.2. Take a piece of card or thick paper and cut in it a slit about one-eighth of an inch wide and long enough to cover the depth of the figure.

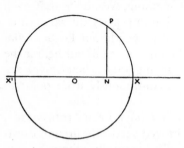

FIG. 1.3.—Simple harmonic motion. As P describes the circle with uniform speed, N executes a simple harmonic vibration along X'OX

Holding the book sideways place the slit at the outer edge of the figure and draw it across the diagram to the inner edge. A disturbance will be seen to start at the left-hand side of the slit and to travel along the slit to the right. The lines across the slit represent the air through which the waves are passing divided into imaginary layers initially of equal width. Watch

PLATE I

FIG. 1.1.—Wave-model. When the handle is turned the balls oscillate vertically and the wave travels horizontally

PLATE II

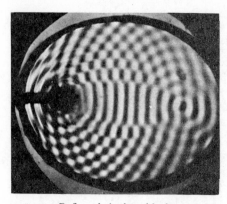

FIG. 1.7.—Reflected ripples with the source at one focus of an elliptical reflector and the reflected ripples converging to the other focus

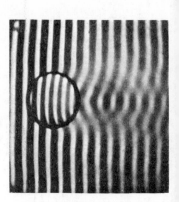

FIG. 1.8.—Straight ripples focused by passing over an immersed lens which gives varying depth, and therefore varying speed for ripples

FIG. 1.9.—Ripples passing through an aperture and showing bending at the edges

FIG. 1.10.—Superposition of two sets of circular ripples showing nodal and antinodal lines. The arrows indicate nodal lines

any one of these layers, and it will be seen to move to and fro along the slit; it therefore represents a longitudinal wave motion. As it moves it alternately expands and contracts, sometimes representing air in a state of rarefaction and sometimes air in a state of compression.

Displacement Curves.—A displacement curve is one the distance of which from the axis at any point represents the displacement of the corresponding part of the medium.   The simplest kind of vibration is that executed by the bob of a pendulum, the prong of a tuning-fork, or a mass suspended by a vertical spring.   Any point in the system moves most rapidly as it passes through its middle position, slows up, reverses, accelerates to a maximum speed as it passes through its middle position in the opposite direction, slows up,

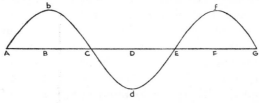

FIG. 1.4.—Displacement diagram for simple harmonic wave. Length of ordinate from axis to curve represents displacement of medium at point for which ordinate is drawn

reverses, accelerates, and repeats the cycle indefinitely.   The exact law of the motion can be studied by thinking of the motion of a point N which is the projection on the line OX of the point P (Fig. 1.3).   If P moves uniformly round the circle and N is the foot of the perpendicular dropped from P on to OX, then N oscillates to and fro in the diameter X'OX of the circle with the type of motion which has been described.   It is called simple harmonic motion and, as will be seen later, it is the type of vibration which in an air-wave gives rise to a pure tone.

Now, in a simple harmonic wave every part of the medium which is transmitting the wave vibrates in this way, and the displacement curve is shown in Fig. 1.4.   Let us suppose that the wave in question is a simple harmonic wave on the surface of the water.   Then the diagram represents the disturbance of the water surface at a particular instant.   ABCDEFG represents the undisturbed level surface, A*b*C*d*E*f*G represents the disturbed surface; *b* and *f* are crests, *d* is a trough.   Waves along a rope generated by moving the end to and fro may be represented in the same way. These are transverse waves—i.e., waves where the particle velocity, the velocity of a part of the medium, is at right angles to the direction of propagation of the waves.

We have just seen, however, that sound-waves are longitudinal —i.e., the motion of the medium is to and fro in the direction of propagation. Suppose that the slit used with Fig. 1.2 represents an imaginary tube along which the sound-waves are being propagated. Each layer is executing a simple harmonic motion, and the same displacement curve (Fig. 1.4) which represents the water-wave will represent the sound-wave if we adopt the simple convention that forward displacements are represented by the distance of the curve *above* the axis and backward displacements by the distance of the curve *below* the axis. Accepting this convention, we see that of the layers which lay undisplaced before the wave disturbed the medium, those lying initially between A and C are all displaced forwards, those initially between C and E are all displaced backwards, and those initially between E and G are all displaced forwards. There is no displacement of the layers A, C, E, and G. There is maximum displacement of the layers initially at B, D, and F.

So far as the density and pressure at various points in the wave are concerned, we may note that for a short distance at *b*, *d*, and *f* the

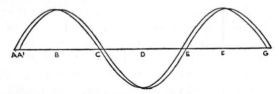

FIG. 1.5.—Two successive positions of a wave to show direction of motion of the medium as the wave passes

curve runs parallel to the axis. Therefore layers initially in the neighbourhood of B, D, and F are all displaced by the same amount. Therefore the layers maintain their normal distance from one another, and the pressure and density remain normal. At C the case is different. The layers just behind are displaced forwards and those in front are displaced backwards. The air at C is thus compressed—its density and pressure are a maximum. At E, on the other hand, since the curve behind E lies below the axis, the layers behind E are displaced backwards, and since the curve in front of E lies above the axis, therefore the corresponding layers are displaced forwards. This leaves E a rarefaction with density a minimum and pressure a minimum.

To see how the particle velocity varies from point to point draw two positions of the wave at a very short interval of time apart, the earlier starting from A and the later from A′ (Fig. 1.5). From A to B the later curve lies below the earlier, therefore forward displacements have decreased and the air is moving backwards. From B to C the later curve lies above the earlier, forward displacements have increased, and the air is moving forward. Since the two curves cut at *b*, the displacement of the layer B has not changed in the interval, and this layer therefore is at rest. Similar considerations show that air layers from C to D are moving forwards and air layers from D to F are moving backwards.

These results can be summarized in the following table:

Layers.	Displacement.	Density.	Pressure.	Particle velocity.
A	Zero	Minimum	Minimum	Maximum backward
A to B	Forward	Below normal	Below normal	Backward
B	Maximum forward	Normal	Normal	Zero
B to C	Forward	Above normal	Above normal	Forward
C	Zero	Maximum	Maximum	Maximum forward
C to D	Backward	Above normal	Above normal	Forward
D	Maximum backward	Normal	Normal	Zero
D to E	Backward	Below normal	Below normal	Backward
E	Zero	Minimum	Minimum	Maximum backward

Fig. 1.6 shows the results graphically, and the motion can be illustrated in detail by using the slit with Fig. 1.2 as previously suggested.

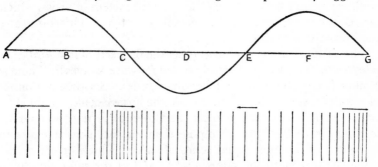

FIG. 1.6.—Relation between displacement, particle velocity, and density. The curve shows the displacement of the medium at a given instant, the arrows show the corresponding particle velocity at the same instant, while the vertical lines indicate the compression and rarefaction of the air

Watching a particular layer from the instant when it is narrowest (maximum compression), we find it moving forward in the direction of the wave with gradually diminishing velocity and expanding at the same time. When it reaches its normal width (normal density) it has no velocity and maximum displacement. It now begins to move backwards, still expanding until it has its greatest width (maximum rarefaction) when it has its maximum backward velocity. Starting now to contract, it continues to move back until its width is normal (density normal), when it stops, reverses the direction of its motion, and, continuing to contract, reaches its maximum forward velocity at the instant of greatest compression.

Properties of Wave-motion.—There are many different kinds of waves, but all these different kinds have certain properties in common. At first sight it is the differences which strike one. What is there in common between water-waves, sound-waves, light-waves, and radio waves? Careful consideration reveals important similarities, and the study of these for familiar kinds of wave-motion is a help to clear thinking about sound-waves. We shall consider four properties common to all kinds of wave-motion—reflection, refraction, superposition or interference, and diffraction. The meaning of these terms will become clear when we consider them in relation to the various types of wave-motion.

Water-waves.—Water-waves provide us with obvious illustrations of some of these properties. Standing on a breakwater or a sea-wall, we cannot help noticing the reflected system of waves which passes outwards through the incoming system. This illustrates the property of *reflection*.

The reflected waves enable us to see the property of *superposition* or *interference*. This was first clearly stated for water-waves by Thomas Young (1773–1829) in the following words: 'Suppose a number of equal waves of water to move upon the surface of a stagnant lake, with a certain constant velocity, and to enter a narrow channel leading out of the lake. Suppose then another similar cause to have excited another equal series of waves, which arrive at the same channel, with the same velocity and at the same time with the first. Neither series of waves will destroy the other, but their effects will be combined; if they enter the channel in such a manner that the elevations of one series coincide with those of the other, they must together produce a series of greater joint elevations; but if the elevations of one series are so situated as to correspond to the depressions of the other, they must exactly fill up those depressions and the surface of the water must remain smooth.' In the case of the direct and reflected waves we have the necessary two series of waves. We see that the reflected waves cross the direct waves without either set losing its identity. We may also notice that at some points the water is always in maximum motion and at some points it is almost at rest. This pattern, which is difficult to observe because the waves are irregular, is steady, and the water surface is said to be in a state of *stationary vibration*. The lines joining points of minimum motion are called nodal lines, and those joining points at which motion is a maximum are antinodal lines.

The observation of *refraction* is not so easy. It occurs where waves cross a boundary separating two media in which they travel with different speed, and it shows itself in a change of direction. There is, in the case of water-waves, no sharp boundary marking a change of speed, and therefore no sharp change in direction for the waves. But where the waves are approaching a shelving beach there is a gradual change of speed, owing to the fact that water-waves travel faster in deep water than in shallow water. If therefore the waves are approaching the beach at an angle, the end of the wave nearer the beach is always travelling slower than the part farther out, and so the waves swing round and tend to become more nearly parallel to the beach. This change of direction is a case of refraction. *Diffraction* may show itself in two ways. We may notice the waves bending round the end of the sea-wall and disturbing the calm water in its lee. We may also notice that when the waves reach a projecting rock, if the part projecting is not too large, the waves tend to bend round it into the ' shadow ' and even, if the projection is small, to re-form behind it and pass on as if no interruption of their progress had taken place. This latter phenomenon occurs only when the obstacle is comparable in size with the wave-length—i.e., with the distance from crest to crest.

Ripples.—The small waves which we call ripples offer still more beautiful illustrations of the properties of wave-motion. Using a strip vibrating on the surface of water in a tank we may produce straight ripples, and using a vibrating point we may produce circular ripples. These may be reflected from straight or from curved boundaries and the resulting pattern photographed. The observations may, of course, be made very simply without apparatus by allowing a drop of water from a wet finger to fall on the surface of the water in one's bath, and watching how the reflected ripples from the side depend on whether the edge of the bath at the point of reflection is straight or curved, and where the point of impact of the water-drop is situated in respect to the reflecting edge. The reflection of ripples originating at one focus of an ellipse and converging after reflection on the other focus is shown in Plate I, Fig. 1.7. The property of refraction must be illustrated by using water of varying depth in the tank. If the water is made just to cover a large lens laid flat on the bottom of the tank, then as ripples cross it they move more slowly over the shallower central portion and more rapidly over the deeper marginal portions, and straight ripples become converging curved ripples focused at a point (Fig. 1.8).

Diffraction can be shown by allowing ripples to fall on a small aperture, when they can be seen to diverge (Fig. 1.9). The narrower the slit, the greater is the divergence. The converse case is seen when ripples fall on an obstacle and bend round into the space at its rear.

Superposition may be observed by using two prongs to generate two sets of diverging ripples. The photograph (Fig. 1.10) clearly shows nodal lines radiating from between the sources. Along these no ripples can be seen, while between these lines strongly marked ripples are apparent—most strongly marked along antinodal lines midway between the nodal lines.

Short Sound-waves.—Coming nearer to our proper subject of study, we can experiment with short sound-waves. Very high-frequency sources produce very short sound-waves, and these illustrate the properties we have been considering. As a source of waves we may use a shrill whistle blown with compressed air, and as a detector we may conveniently use a sensitive flame. This consists of a tall flame, ordinarily quite steady, produced by forcing gas at fairly high pressure through a fine nozzle. If the pressure is too high the flame drops and roars. As the pressure is lowered a point is reached when the flame is tall and steady, but very sensitive to disturbances in the air. If it is exposed to a draught it at once drops and roars, and the same effect is produced by the sounds due to the sibilants of speech, the shaking of keys, the dropping of a coin, and, of course, the blowing of the whistle. The property of reflection may be shown by placing two concave mirrors some distance apart with the whistle at the focus of one and the sensitive flame at the focus of the other.

If now the two mirrors are arranged so that their two axes are
in line, the flame is strongly affected; but if the mirror which is
reflecting the waves from the whistle is slewed round slightly
so as to direct the beam of sound to one side of the other mirror,
the flame remains steady.

Superposition can be shown by attaching to the whistle a
T-shaped tube the stem of which receives the whistle, while the
cross-piece carries apertures at its two ends. The waves emerging
from the two apertures pass through one another, and in doing so
form a system of surfaces where the effects of the two waves are
opposed (nodal surfaces) and surfaces where the effects of the two
waves are added (antinodal surfaces). If the flame is moved
about, it will demonstrate the existence and position of these
surfaces by responding vigorously to the motion of the air at
the antinodal surfaces and remaining entirely quiescent at the
nodal surfaces. As any motion of the flame is apt to disturb it,
the effect is shown more convincingly by leaving the flame in
position and moving round the whistle and T-tube. This movement carries the nodal and antinodal surfaces with it, and the
flame varies accordingly.

The simplest type of superposition occurs when two equal sets
of waves cross one another moving in opposite directions. If we
make a flat plate oscillate to and fro, longitudinal air-waves
are produced. In moving forward it compresses the layer of air
immediately in front of it. This layer relieves its compression
by expanding into the layer in front and compressing that, and so
the compression is propagated from point to point through the
medium. Meantime, when the plate moves back it leaves the air
immediately in front rarefied. The air beyond this, being still
at normal pressure, expands back into the rarefied layer, itself
becoming rarefied, and so a rarefaction moves outward from the
oscillating plate. Thus the air over any plane parallel to the
plate moves backward and forward, while across it there travels
a series of alternate equidistant compressions and rarefactions,
as illustrated by the slit used with Fig. 1.2. If now, opposite
to the oscillating plate and parallel to it we place a board, the
waves will be reflected back from the board, and will be superposed on the oncoming waves.

The result of the superposition is shown in Fig. 1.11. The direct
wave is travelling from left to right, and is represented by the broken
line. The reflected wave is travelling from right to left, and is repre-

sented by the broken line with dots. On the top line the two sets of
waves are in phase, and the result of the superposition is given by the
continuous line, and shows that the air at A, C, E, and G is undisplaced,
while that at B and F has a double displacement forwards and that at
D a double displacement backwards.   The next line shows the position
after the direct and reflected wave systems have each travelled through

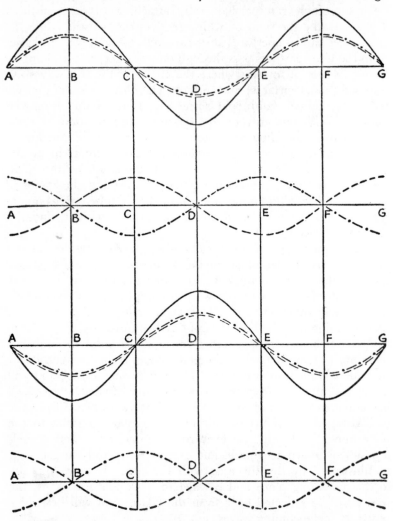

FIG. 1.11.—Stationary waves due to two equal sets of waves travelling in opposite
   directions. The broken line with dots represents waves travelling to the left.
   The broken line without dots represents waves travelling to the right. Four successive instants are shown

one-quarter of a wave-length. The two waves are now in exactly opposite phase. At C, for instance, the direct wave alone would produce a maximum displacement forwards (since the broken curve lies above the axis), while the reflected wave alone would produce a maximum displacement backwards, the corresponding curve being below the axis. At every point the displacements which the two waves would severally produce would be equal and opposite. At this particular instant, therefore, the medium is completely undisplaced. In the third line of the diagram we see the position after both waves have moved through a further quarter of a wave-length. At B and F there

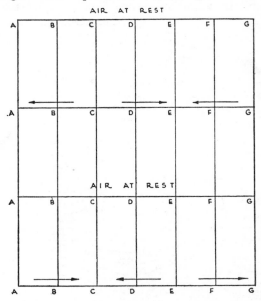

FIG. 1.12.—Motion of the air at four successive instants separated by one quarter of a period

is now a double displacement backwards, at D a double displacement forwards, while A, C, E, and G remain undisplaced. In the next stage, shown in the fourth line of the figure, the two waves are in opposition and the whole medium is once more undisplaced. The cycle of changes then repeats itself. Throughout the cycle the air at A, C, E, and G remains undisplaced. That at B, D, and F is oscillating between a double forward displacement and a double backward displacement.

Fig. 1.12 shows how the air is moving at the four instants represented in Fig. 1.11. When a medium is transmitting one set of waves every point in the medium executes the same vibration in a different phase—i.e., it is just a little earlier in its stage of

vibration than the points farther from the source and just a little
later than those nearer the source. In the *stationary vibration*
which results from the transmission of two equal waves in opposite
directions this is not so. There are points of no motion (nodes),
A, C, E, G, and points of maximum motion (antinodes), B, D, F.
All the points between A and C are in phase with B—i.e., move
in the same direction at the same time and reverse their motion at
the same time. They only differ among themselves in amplitude,
the extent of the vibration getting smaller as we move from B

FIG. 1.13.—Wave diagram to illustrate stationary waves due to the superposition of
two equal sets of waves moving in opposite directions through the same medium

towards either A or C. The same is true of the relation between
the motion of all points between C and E and that of D, and
again of the relation between the motion of all points between
E and G and that of F. It is to be noted also that the movement
of the air on the opposite sides of a node is always towards
the node in both cases, or away from it in both cases. Lastly—
and this is very important—the distance from a node to the
next antinode is one-quarter of a wave-length and the distance
between two successive nodes or two successive antinodes is one-
half of a wave-length. If the slit used with Fig. 1.2 is now used
with Fig. 1.13 this motion can be studied in detail.

The existence of these nodal and antinodal points can be demonstrated by placing the sensitive flame between the whistle and a board which reflects the waves back again towards the whistle. The air between the whistle and the board is divided up into nodal and antinodal layers. If the flame is moved—or better, if the board is moved backward or forward—the flame responds strongly or is quiescent, according as the direct and reflected waves at the point where it is situated are in phase or out of phase. If the board is drawn steadily back the series of nodes and antinodes moves with it and the flame records their passage. Thus it is easy to find the distance between two nodes, and therefore the wave-length and frequency of the note of the whistle.

Thus (p. 3) if c is the velocity of sound, f the frequency, and λ the wave-length

$$c = f\lambda \quad \therefore f = \frac{c}{\lambda}$$

If $c = 340$ m/s (the value at about $15°$ C.)
and $\lambda = 340$ mm (a very high-pitched note)
$f = 10{,}000$ Hz.

Hertz, abbreviated Hz, is the unit of frequency and is defined as the number of cycles per second.

All sounding bodies are in a state of stationary vibration. In the vibrating string waves are passing to and fro reflected at the two fixed ends, the two systems combining to give the actual observed vibrations. In the organ pipe, and all wind instruments, waves pass from one end of the air column to the other and back, reflected from the ends and superposing their effects to give the actual vibration of the air in the pipe. If a plate is hit at one point the resulting note is due to the superposition of waves starting at the point of impact, travelling through the plate, and combining with the waves reflected at the edges and travelling back towards the point of impact.

Diffraction can be shown by placing obstacles between the whistle and the flame and showing that for small obstacles the sound bends round into the ' shadow ' and affects the flame. The most striking demonstration is the exploration with a sensitive flame of the sound ' shadow ' of a circular disc. If the whistle and the flame are placed exactly on the axis of the disc, then the flame responds almost as strongly as when the disc is removed. A slight displacement of the disc destroys the effect

at once, showing that it is only the centre of the shadow which is
so affected. This experiment is the exact analogue of that used
to illustrate Fresnel's theory of the diffraction of light. If a
strongly illuminated pinhole be used to throw a shadow of a
small, perfectly circular disc (e.g., a new halfpenny piece) on a
distant screen a bright spot will be found at the centre of the
shadow.

Ordinary Sound-waves.—Short sound-waves are, as we have
seen, produced by high-pitched notes, and since they are short
—their wave-length is only a few centimetres—the apparatus
necessary for demonstrating their properties can be arranged on a
similar small scale. Ordinary sound-waves are much longer,
however, and in the case of the fundamental pitch of the male
voice are about $2\frac{1}{2}$ to 3 metres wave-length. In this case we
cannot experiment on the waves on a small scale, and must
use methods of large-scale observation. Reflection of ordinary
sound-waves is familiar to us in the case of the echo. This is
caused by the reflection of the sound-waves from a cliff, or from
the gable end of a house, or sometimes from the edge of a wood.
In a valley or glen with rocky sides a multiple reflection can sometimes be heard, the later echoes coming from more remote reflecting
surfaces. Sometimes an echo may be heard in a large hall,
and it may be a very troublesome feature if the hall is used
for public speech or music. It is most troublesome when it is
associated with a curved surface, and walls which are circular in
plan, or ceilings which are dome or barrel-vault in shape, have to
be very carefully designed to prevent the reflected sound being
focused at particular points and giving prominent and well-
marked echoes (see also p. 241). Troublesome echoes are rarely
found where all the surfaces are plane. Just as polished metal
reflects light copiously and dull black surfaces hardly reflect
it at all, so surfaces vary very greatly in the proportion of sound
which they reflect. Smooth, hard surfaces have a very high
reflecting power, soft, porous surfaces reflect very little sound and
are said to be ' good absorbers '. These absorbers play a very
important part in the acoustic design of buildings (p. 230).

Another interesting case of reflection is the echo from the sea-
bed. The old method of sounding involves the use of a line and
sinker, and in order to take a sounding the speed of the ship must
be considerably reduced. Taking soundings by this method in
very deep water is a laborious business, involving an elaborate

engineering outfit. Soundings can be taken quickly and conveniently, however, by echo methods. A sound is made just under the surface of the water. This is transmitted to the sea-bed, where it is reflected and reaches the surface again. If we know the speed of sound in water and the total time taken for the sound to reach the bottom and the echo to return, then we can easily calculate the depth of the water. In the devices now available we are spared the trouble of calculation, and depths can be read off directly on a dial. The method is now extensively used.

Refraction is a property of audible sounds which is frequently observed without being understood. It is well known that sounds ' carry ' badly in the middle of a hot day and carry very much better in the cool of the evening. This is mainly due to refraction. Sound-waves travel more rapidly in warm air than in cool air. At midday the ground becomes hot and heats the layers of air in contact with it. As we rise in the atmosphere the temperature falls. Thus when sound-waves are travelling along near the ground, the lower edge of the wave, being in warmer air, travels faster than the upper edge, and the waves are deviated upwards in consequence and tend to miss observers on the ground. On the other hand, in the evening the ground cools quicker than the atmosphere, and cools the layers of air in contact with it, so that the lower layers are cooler than the upper layers. In this case the upper edges of the sound-waves travel faster than the lower edges, and the waves are bent over towards the earth and keep along the ground.

A similar effect is observed in the case of wind, but is not such a clear example of refraction. It is a well-known fact that sounds carry well with the wind and badly against it. This difference is not due simply to wind velocity, but to wind-velocity *gradient*. If the wind blew with the same velocity at all heights it would have practically no effect on the audibility of distant sounds. This, however, is not the case. Near the earth's surface wind velocity increases with height. This variation is the wind-velocity *gradient*. It follows that for sound-waves proceeding in the direction of the wind the top edges get blown forward more than the bottom edges, and the waves are directed towards the earth and kept along the ground. For sounds travelling against the wind the top edges are retarded more than the lower edges, and the waves are deviated so as to leave the surface of the earth and

become inaudible to observers on it. The inaudibility is not
due to the sound-waves being dissipated; the sound is still very
loud, but in order to hear it the observer would have to move to
some height above ground level.

A specially interesting phenomenon occurs when the sound-
waves deviated upwards by a temperature or wind-velocity
gradient meet a *reversed* gradient—e.g., a layer where the wind
velocity begins to get less or the temperature begins to diminish
as the height increases. This kind of effect may occur both for
wind and for temperature. When it happens, the waves are again
deviated downwards, and return to the earth to be audible on
the other side of a ' silent zone ' (Fig. 1.14). This effect is not at
all uncommon, and was noted in the case of a naval engagement

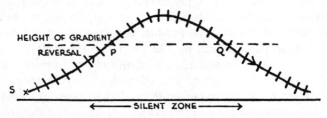

FIG. 1.14.—Refraction of sound-rays due to reversal of wind or temperature gradient
which causes them to leave the surface of the earth near the source and return
to the surface at a more distant point

with the Dutch fleet in 1666 both by Pepys and by Evelyn in
their respective diaries. Modern instances are numerous.

We have seen that superposition involves the simultaneous
transmission of independent series of waves by the same medium.
This is also the condition of the enjoyment of good music. In
quartet-playing the waves from each instrument traverse the air
in the room without losing individuality, and from the mass of
tone the ear can select and follow any one part. Similarly in
orchestral music from the still more complex mass of tone the
strings, woodwinds, brasses, &c., can be separated out, and even
the waves from solo instruments can be isolated by an act of
attention and followed at will.

The interference effects of superposition are less commonly
observed. This is partly due to two reasons. Most ordinary
sounds are mixed—i.e., they contain sounds of different wave-
lengths. Now, the distribution of maxima and minima due to
superposition depends on the wave-length and varies with the

wave-length. Thus a point which is a maximum for some constituents of the sound may be a minimum for others and vice versa. This tends to even out the loudness of the sound. Then, again, we have two ears, and if one of them is at a point of maximum loudness the other is some distance from the maximum—for short waves may even be at a minimum—and this, again, tends to average the effects of superposition. When a high-pitched whistle is sounded in a room the ' superposition pattern ' is very marked, especially if the note is nearly pure. If one ear is closed with the finger and the head moved about, very marked changes of loudness are noticed.

When a gramophone or loud-speaker is being tested for the loudness which it gives for various notes, this phenomenon of superposition causes serious inconvenience. The sound is picked up by a microphone and measured electrically, but everything depends on the placing of the microphone. If this is at a

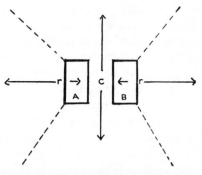

FIG. 1.15.—Diagram of tuning-fork showing compression, C, and rarefactions, *r*, travelling out from the prongs, A, B, of a tuning-fork as they approach. Along the dotted lines compressions and rarefactions are superposed and no sound is heard

maximum of the superposition pattern it will give one result, if at a minimum it will give an entirely different one. In order that the average effect may be measured, the microphone may be made to swing as the bob of a long pendulum so that it moves to and fro through several feet and includes in its passage maxima and minima. The effect, which is of course due to the superposition of reflected waves from wall, ceilings, &c., is so persistent that no amount of lagging on the walls will deaden the reflection sufficiently to eliminate its effect completely.

When a tuning-fork is rotated about its shaft in front of the ear, alternations of sound and silence are heard. Four times in every complete revolution the sound is a maximum and four times a minimum. This may be regarded as due to the fact that when the prongs are approaching one another rarefactions start from behind the prongs and a compression from between them, as shown in Fig. 1.15. Along four lines these will be superposed so as to oppose one another. When the prongs are moving apart the conditions as to compressions and rarefactions are reversed, but it is still true that in the directions shown these are superposed so as to oppose one another. Along these lines, then, no changes of pressure are propagated, and in these directions no sound is heard.

Another very important instance of superposition is the phenomenon of *beats*. When two sources of sound of nearly the same frequency are sounded simultaneously, the resulting effect is a periodic alternation of sound and silence which we call beats. It is easy to see how the effect arises. At a given instant the two sets of waves superposed at the observer's ear are in agreement. Compressions arrive together and rarefactions arrive together, and we hear a loud sound. One of the sources, however, is vibrating slightly faster than the other. The waves it sends out begin to arrive earlier than those from the other source. Soon the faster source has gained half a vibration on the other. The superposed waves are now in opposition. A compression from one source arrives simultaneously with a rarefaction from the other and vice versa. If the two sounds have equal intensity there will be complete silence. In any case there will be a minimum of sound. Presently the faster source will have gained a whole vibration, the two sets of waves will be in agreement again at the ear, and the sound will give maximum loudness. Clearly the number of beats per second will be the number of complete vibrations which the faster source gains on the slower in one second, and this in turn will be the difference in frequency of the two sources.

Beats can very easily be illustrated experimentally. If two tuning-forks in unison are taken, and the prongs of one of them loaded slightly with soft wax, the pitch of its note will be flattened, and when the two are sounded together, beats will be very distinctly heard. They are most distinct when the two sources have equal strength. The phenomenon of beats has been applied in a very interesting way to the detection of gas in mines. Two identical whistles are blown simultaneously—one with pure compressed air from a cylinder, the other with the air in the mine. So long as this latter air is pure the two whistles are in unison, but a very slight admixture of lighter gases causes a rise in pitch of the second whistle and very noticeable beating. The proportion of impurity present can be estimated from the rate of beating.

Beats are very important in tuning two sources to unison.

They get slower as unison is approached, and disappear when unison is reached. They can also be used to adjust two notes to any difference of frequency which does not involve beats too rapid to be counted. Their application in certain organ-stops, and their importance in the theory of dissonance will be considered later.

Diffraction is so familiar that it hardly requires pointing out. Sounds are often heard when the source is invisible, and this is usually due to the fact that the sound-waves spread round the obstacles in their path. On the other hand, sound may enter a room through a chink of open window and diverge so as to spread through the whole room.

Since, then, the transmission of sound is accompanied by the same phenomena as we find associated with all kinds of wave-motion, it is legitimate to regard sound as being propagated by waves, the waves being longitudinal in type.

Transmission of Sound.—Sound-waves travel through any material medium—gas, liquid, or solid. In air they travel with a velocity which varies with the temperature, but at 15° C. (59° F.) it is 1120 feet per second, or about 340 metres per second. It is this finite velocity of sound which accounts for the lag between seeing the smoke of a distant gun and hearing the report. Light travels almost instantaneously. In the same way we can estimate the distance of a lightning discharge by timing the interval between the flash and the report and remembering that the sound takes about five seconds to travel one mile. Sound-waves also travel in liquids. Immerse your ears under water and you will hear distinctly sounds that are produced in the water. Signals can be transmitted over very long distances under water and are subject to much less disturbance than in air. The velocity of sound in water is about 1420 metres per second, or more than four times its velocity in air. Lastly, sound-waves can travel in solids. The velocity of sound through a solid bar is much greater than that in either air or water, and may be as much as 3000 or 4000 metres per second. The phrase ' having one's ear to the ground ' probably originated in the habit attributed to American Indians, among others, of listening for the sound of a distant rider on horseback, as the impact of the hoofs on the ground is transmitted through the earth and picked up by an ear pressed to the ground. The transmission of sounds from room to room in a house or a set of flats can be very troublesome, and is mainly due to sounds which are ' structure borne '—i.e., transmitted by walls, floors, ceilings, &c.—as opposed to ' air borne '—i.e., transmitted through the air and through cracks, chinks, &c.   These structure-borne sounds are always worst when the noise is produced by an impact on the structure—e.g., a footfall—or by a source of sound in direct and close contact with the structure—e.g., a piano or wireless set standing on the bare floor without a mat or carpet.

CHAPTER 2

FORCED VIBRATION AND RESONANCE

Free Vibration.—Any source of sound if set in vibration and left to itself vibrates in its own natural frequency, producing a note which gradually dies away as the vibrations decrease, but remains constant in pitch. This type of vibration is called 'free vibration'. A tuning-fork struck and left to itself, a stretched string plucked or bowed and left to itself are obvious examples. The rate at which the vibrations die out varies a good deal. In the case of a tuning-fork the vibrations last for a long time; in the case of the stretched string they do not, as a rule, last so long; in the case of an air cavity they die out very rapidly. Produce a note by blowing across the mouth of a bottle. The note is due to the air contained in the bottle being set in vibration. Immediately the air-blast ceases the note seems to cease. The vibrations die out with extreme rapidity. In these cases we say that the difference is one of damping—that the vibrations of the tuning-fork are only lightly damped, while those of the air cavity are heavily damped. When a tuning-fork is struck a certain amount of energy is communicated to it, and this energy appears in the vibrations of the prongs. This energy is used up in two ways. Some of it is dissipated in overcoming the resistance of the air to the movement of the prongs through it. This energy appears as heat, just as does any energy which is used up in overcoming friction—e.g., in the bearing of a wheel. Some of the energy, however, is communicated to the air as sound-waves and conveyed through the air away from the fork. If either of these sources of energy-loss is increased the rate of damping increases also. For instance, if a large tuning-fork be excited by a blow and held in the air, its rate of damping is very slow, but if it be given a blow of the same strength and then placed with its shaft on a table or on a resonance box (see p. 26), the sound given by the fork is much louder and the rate of damping is increased so as to shorten appreciably the duration of the sound. Sound-waves are carrying energy more rapidly away from the fork, and the vibrations therefore decay more rapidly.

23

Forced Vibration.—If a force which varies periodically is applied to a vibrating system, the system vibrates in the period of the force with an amplitude which is generally small. This is forced vibration. For instance, if a sounding tuning-fork is held over the mouth of a bottle there will generally be a feeble sound from the air in the bottle, and the sound will have the pitch of the fork. Blowing across the mouth of the bottle will elicit the usually quite different pitch corresponding to the free vibrations of the air in the bottle. In the previous case the vibration was forced, the air-waves from the tuning-fork being the periodic force. When a violin is played, the sound comes mainly from the body of the instrument and partly from the contained air. These vibrating systems give the pitch of the note of the string. They are in forced vibration. The body of the violin and the contained air both have a tone of their own; but their vibrations are not free, they are controlled by the vibrating string. Musical instruments generally provide examples of this kind of vibration in so far as they consist of two vibrating systems, one of which imposes its vibrations on the other. It is this second system which, as a rule, radiates the sound. In all the bowed-string instruments the conditions are similar. The 'cello string forces the vibration of the body of the instrument. In the piano the string controls the vibrations of the sound-board from which the sound is radiated. In the case of wind instruments with reeds—e.g., oboe, clarinet, and bassoon—and instruments where the performer's lips act as reeds—e.g., the horn and the cornet— we have the reeds acting as the periodic force, and the air-column in the instrument forced into vibration more or less in the period of the reed and radiating the sound.

Strictly speaking, in forced vibration the force acts on the system without any appreciable reaction of the system on the force. This means that, while the force determines the period of the system on which it acts, the system has no effect on the period of the force. If the piano is an illustration of forced vibration, the vibrating string must impose its frequency on the sound-board without the frequency of the string being in any way modified by the sound-board. This is approximately true. We can see the conditions involved if we think of two pendulums of unequal length suspended from a horizontal string (Fig. 2.1). They form a coupled system. If one of them is set in motion, the impulses communicated along the connecting string set the other in motion. The result is a complicated type of vibration, unless the conditions for forced vibration are nearly fulfilled. These will be found to be twofold: (1) the driving pendulum must be much more massive than the driven pendulum, (2) the coupling must be tight. The first condition is achieved if one pendulum is of lead and the other of wood or some light material. The second condition is that the two points of suspension should be fairly close together. Obviously the farther apart the points of suspension the less is the influence of one pendulum on the other. If these conditions are fulfilled and the two pendulums are of different lengths, and therefore have different natural frequencies of vibration, we shall find that the lighter pendulum will be compelled to vibrate with a frequency which is nearly the natural frequency of the heavy one if this latter is held to one side and then released.

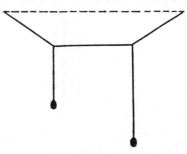

FIG. 2.1.—' Coupled ' pendulums suspended so that the motion of one is communicated to the other

A good example of a coupled system which approximates to a case of forced vibration is provided by a reed and associated air column. If the natural frequency of the reed and the natural frequency of the air column are different, then when the combination is blown the frequency of the note which is radiated from the air column is not the natural frequency of the air column, nor yet the natural frequency of the reed imposed on the air column, but a new compromise frequency which results from the interaction of the two. The instrument is so designed that this new frequency differs very little from the natural frequency of the air column, the reed being pulled out of its natural frequency so as to vibrate nearly with the frequency of the air column. This is achieved by having the air column large and its coupling tight—e.g., without a leak between the air column and the reed due to a badly fitting mouthpiece. It will be seen later that the flute and the flue-pipes of the organ are also examples of coupled systems.

Resonance.—The principle underlying the phenomenon of resonance was clearly stated by Galileo (1564–1642). He

remarks that every pendulum has its own definite and deter-
minate period of vibration which nature has given it, and
cannot be induced to vibrate in any other. On the other hand,
even a heavy pendulum can be set in vibration by the breath
if we blow upon it intermittently and keep time with its
swings. He goes on, 'Even as a boy, I observed that one man
alone, by giving impulses at the right instant, was able to ring
a bell so large that when four, or even six, men seized the rope

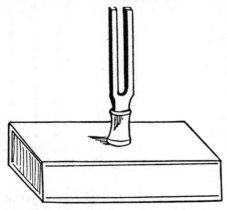

FIG. 2.2.—Tuning-fork mounted on resonance box, the air in which radiates the note of the fork

and tried to stop it they
were lifted from the
ground, all of them to-
gether being unable to
counterbalance the mo-
mentum which a single
man, by properly-timed
pulls, had given it '.
Wallis (1616–1703) [1] contributes a paper ' On the
Trembling of Consonant
Strings ', calling atten-
tion to the fact that if a
viol or lute be touched
with the bow or hand,
another string on the
same or another instrument not far from it, if in unison to it
or an octave, or the like, will at the same time tremble of
its own accord. He adds, ' I have heard of a thin fine Venice
glass cracked with the strong and lasting sound of a trumpet or
cornet near it, sounding a unison or consonant note to that of
the glass. And I do not judge the thing very unlikely.' The
response of a sounding body when its own natural note is played
or, in wider terms, the response of a vibrating system when
subjected to a force timed to its own period is called ' resonance '.
Thus resonance is the particular case of forced vibration when
the force and the system are in unison. Fill a bottle with water
until the pitch of the note elicited by blowing across the mouth
is *g*. Then take a series of forks from *c* to *c'*, strike each one in
turn, and hold it over the mouth of the bottle. The response
will be small for *c* and *c'* but greater for notes nearer to *g*, and
greatest of all for *g* itself. The air in the bottle is said to respond

[1] *Phil. Trans.*, Vol. 12, p. 380 (1677).

by resonance. A rectangular wooden box of suitable size with one end open may be constructed so that the natural frequency of the contained air is the same as that of a given fork. If now the fork is mounted on the top of the box, the air is forced into vibration, and the pitch of the fork and of the air being the same, we get resonance, and the sound of the fork being radiated from the air cavity is greatly increased. Such a device, due to Marloye, is known as a resonance box (Fig. 2.2). Helmholtz (1821–1894) in his experiments on the analysis of sounds used a series of air resonators shown in Fig. 2.3 (p. 62). Experiments with the bottle show that the pitch of the note rises as the volume of air contained diminishes. This is very noticeable as we fill a jug or bottle at a tap. It will also be found that the pitch drops as the aperture is diminished—e.g., by drawing a card partially over the mouth. It is therefore possible to tune an air resonator by altering either or both of these factors. In this way Helmholtz tuned the series of resonators shown.

Sharpness of Resonance.—We have seen that if a series of tuning-forks is held successively over the air in a bottle the response is greatest to the fork whose pitch is that of the air in the bottle. But if we try the experiment out carefully we shall find that the resonance is not sharp—i.e., we not only get a response to the correctly tuned fork, but we get a response, less marked it is true, but quite appreciable, to forks a semitone, a tone, or even a third or fourth from the correct pitch. Thus the response does not vary greatly with the correctness of the tuning, and is still noticeable when the mistuning is considerable Let us compare this with another case. Take two tuning-forks of the same frequency mounted on resonance boxes and place them on the table with the open ends facing one another. If one fork is struck and then stopped with the finger, the other will be heard sounding in resonance. If, after a few seconds, this in turn be stopped, the first will now be found to be sounding again, and in this way the energy may be transferred from fork to fork several times. If now one of the forks be loaded by attaching a small piece of wax to the end of one prong, its pitch will be lowered. The lowering may be too small to be detected by ear, and yet if the experiment be repeated it will be found that there is almost no response even in the first instance, and transfer of energy from fork to fork is impossible. In the case of the air in the bottle we get response over a wide range of pitch without very marked response at correct tuning. In the case of the tuning-fork we get marked response at correct tuning and almost no response for incorrect tuning, even if the error is quite small. In the first case the resonance is said to be general, in the second case it is highly selective. Those interested in radio will see an analogy here. A cheap radio set responds over a large range of wave-lengths without marked response, even when correctly tuned. A good set is selective—it responds strongly to the waves to which it is accurately tuned and cuts out—i.e., gives no response at all—to neighbouring wave-lengths. In the case of sound a vibrating wire stretched between bridges is intermediate between the tuning-fork and the air cavity; it is less selective than the tuning-fork, but more selective than the air cavity. It is found that selectivity depends on damping. In heavily damped systems like the air cavity the selectivity is small, while in lightly damped systems like the tuning-fork the selectivity is high.

This is an important point when we are considering the possibilities of a 'radiator' for sound. A radiator which transmitted one note at strong intensity and others feebly would be very unsatisfactory. Even an air column of fixed length would be unsatisfactory because of its fairly marked response to notes of its own proper pitch. That is why when we use air columns (as in the organ pipe, trombone, cornet, &c.) we use a different length of pipe for each note or a different partial of the pipe (see p. 111). The sound-board of a piano, on the other hand, gives a general resonance without any selective response. This can be achieved in one of two ways. We can, for instance, arrange that the natural frequency of the system is outside the range over which a general response is required. If the diaphragm of a telephone transmitter had a natural frequency within the range of frequencies it was required to pick up and transmit, it would distort all speech by responding selectively to its own pitch. If, on the other hand, we arrange for it to have a frequency of 6000 or more, it will not be very sensitive at lower frequencies, but its response will be fairly uniform. Another way in which the general response can be obtained is to arrange that the responding system has a series of overlapping regions of pitch in which it gives resonance so that the resonance is distributed.

Combination Tones and Harmonic Partials.—If we listen carefully to the sound of a referee's whistle (Fig. 2.4) which is strongly blown, we shall find that if, as is generally the case, it consists of two barrels, the resulting sound is more complex than we should expect. The two high-pitched notes due to the two barrels of the whistle are clearly present. But in addition, and extremely prominent when attention has been drawn to it, there is a comparatively low-pitched buzz which gives character to the whole sound. It is found that the frequency of this low-pitched buzz is the difference of the frequencies of the two high-pitched notes.

The phenomenon seems to have been discovered originally by the violinist Tartini (1692–1770). In his *Trattato de Musica*, published at Padua in 1754, he refers to ' terzi suoni ' (third sounds) and gives examples. In a later work, *De' Principi dell' Armonia Musicale*, published at Padua in 1767, he refers again in the preface to the discovery of third sounds, and in his second chapter says [1] that ' in the year 1714, when

FIG. 2.4. — Referee's whistles. Showing two barrels giving strong low pitch difference tone

a youth of about twenty-two years, he (Tartini) accidentally discovered this phenomenon on the violin in Ancona, where not a few who remember the fact are still living. He immediately began communicating the discovery to professors of the violin without reserve or mystery. He made it the fundamental rule of perfect tuning for the boys in the school which he opened in 1728 at Padua, and which still exists. And thus the knowledge of the phenomenon spread throughout Europe.'

Sorge, a German organist, is sometimes credited with the discovery of these tones, but according to the article just quoted he refers to them in a book published in 1745, but nowhere claims to have discovered them. There is more solid evidence of an independent but later discovery (1751) of these tones by Jean Baptiste Romieu.

The tone being the result of the simultaneous sounding of two other tones is called a ' combination tone ', and because its frequency is the difference of the frequencies of the two generating tones, it is called a ' difference ' tone. We shall see presently that there are other combination tones. The difference tone is the most prominent, and as a general rule it is easiest to hear

[1] A. T. Jones, *American Physics Teacher*, Vol. 3, p. 49 (1935).

it if the two generating tones are high-pitched and so chosen that the difference tone comes in the range of pitch to which the ear is most sensitive—i.e., about 3,000 Hz. It is particularly well heard on a loud-toned harmonium, but with practice may be heard with the piano, and especially with the violin. It must obviously be taken into account in considering the complex effect produced when two notes are sounded together.

Since the frequency of this tone is the difference of frequency of the generating tones, we might regard it as being due to the coalescence of rapid beats to give the impression of a tone. There are strong reasons against this view, however. Beats are a periodic lulling and swelling of a tone whose pitch lies between those of the generating tones. The combination tone is a pure tone of entirely different pitch, and may be either well above or well below the pitches of both generating tones. Helmholtz was the first to appreciate the real importance of these tones, and investigated their origin with the thoroughness which characterized all his work and with that knowledge of physiology, physics, and music which enabled him to revolutionize the subject of musical acoustics and produce his classical work *Sensations of Tone*.[1]

He looked elsewhere for the explanation of combination tones. Most vibrating systems, if the amplitude is very small, give approximately simple harmonic vibrations when vibrating freely. This means that for small amplitudes the force that restores the system when displaced is proportional to the distance through which it is displaced. For instance, if we pull a violin string to one side through a small distance by exercising a certain force, it will take twice the force to displace it twice the distance. In so far as this relationship holds, the free vibrations of the system are simple harmonic. But in point of fact, for some systems this relationship does not hold at all, while for others it holds only for very small displacements, and probably in no case of a sounding body does it hold at the displacements necessary for moderately loud sounds. Helmholtz showed that if a system for which the relationship holds approximately is acted upon simultaneously by two periodic forces, then its vibrations show the frequencies of the two forces only as we should expect. If, however, a more complicated relationship holds, and the system

[1] Translated by A. J. Ellis and first published by Longmans and Company in 1875. Republished as a paperback by Dover Publications, Inc.

is a ' non-linear vibrator ', then in addition to the two expected frequencies others appear. If, for instance, the frequencies of the two forces are f_1, f_2, then the system on which they act gives not only f_1 and f_2, but in addition a frequency $f_1 - f_2$—that of the difference tone. The mathematical analysis predicts other tones, particularly the frequencies $2f_1$, $2f_2$ and $f_1 + f_2$. The first two are harmonic partials (see Chap. 5) of the generating tones. The last is another combination tone—the ' summation ' tone, so called because its frequency is the sum of the frequencies of the two generating tones. It is much more difficult to hear than the difference tone, the best conditions being given by the harmonium when we sound strongly two notes of low pitch, chosen so that their summation tone occurs in the region of pitch to which the ear is most sensitive. This summation tone cannot be explained by beats, and in any case Helmholtz showed that these combination tones are a natural consequence of the double forcing of the kind of system frequently met with in sounding bodies. More important still, we have good reason to believe that parts of the mechanism of the ear behave in the required way even for very small amplitudes, so that combination tones, even if not produced externally to the ear at all, may be produced in the ear itself, and the perceived sound modified in consequence. The evidence for the production of these aural combination tones will be considered in Chapter 6. Of course, if only one generating tone is present the combination tones disappear, but the harmonic partial of that particular tone remains, and the production of these tones in the ear will also be considered later.

In some instances the phenomenon is still more complicated, either in the case of louder generating tones or where the system acted on by the generating tones departs in a more marked way from the conditions necessary for simple harmonic motion. In these instances each of the combination tones so far considered (the ' first-order ' combination tones) can act as a generating tone and combine with one of the original generators to form a ' second-order ' tone. Thus we can have $f_1 - 2f_2$, $2f_1 - f_2$. We can also have the third-order partial tones $3f_1$, $3f_2$. It is difficult to estimate the practical importance of the second-order combination tones, but third-order partial tones are often produced in the ear with sufficient intensity to be perceptible.

Combination tones are important in several different ways.

(1) The production of the difference tones in the ear may explain the way in which the ear supplies a missing fundamental or strengthens a very weak one (p. 86).   (2) The difference tone is used on some organs to produce an ' acoustic bass ' (p. 129). In the interests of economy the difference tone between the notes of two short pipes may be used instead of the note produced by a longer pipe.   (3) The harmoniousness of a diad or triad may be affected by the combination tones (p. 165).   (4) The fact that in the old-fashioned gramophone with the small horn a bass was heard which the horn could not possibly transmit is probably due to the combination tones produced in the ear filling the necessary gap.

CHAPTER 3

INTENSITY AND LOUDNESS

Power of Sources of Sound.—When the steam engine was first invented it had to replace the horse as a source of power. The pumping of water from the mines was done by pumps worked by horses. Before the mine-owner would introduce the new invention he had to know how many horses it would replace, and so be able to compare fuel costs with fodder costs. The horse-power was originally the *rate* at which an average horse could do work over an average working day. It is now fixed in terms of the rate at which a weight can be lifted, and is the power required to lift 550 lbs. one foot per second. The watt is approximately $\frac{1}{746}$ of a horse-power. The engine of a motor-car is generally rated in horse-power. The large dynamo is rated in kilowatts, the kilowatt being 1000 watts, or about $1\frac{1}{3}$ horse-power. Very small sources of power are rated in microwatts, the microwatt being one millionth of a watt. Some idea of the magnitudes involved in these units will be possible if we note that 1 watt is the power required to raise a weight of 1 lb. steadily by about $8\frac{1}{2}$ inches per second; that a man doing hard continuous manual labour develops a power of about 100 watts; and that this is in turn the power required to keep alight a 100-watt electric lamp.

When we come to consider the power of a source of sound, however, we are not thinking of the power required to drive the source and maintain it in vibration, but of the power actually radiated as sound. These two things may be very different. The musical instrument is a ' transformer '. It is supplied with power by an engine or a performer. Much of this power is used to overcome friction and is wasted in other ways. Only a small proportion is transformed into audible sound. Thus a large organ may require an engine developing 10 kilowatts (10,000 watts) to blow it, yet all that appears as sound may be 12–14 watts. A pianist may use energy at the rate of 200 watts in a very loud passage without more than about 0·4 watt being radiated as sound. The human voice is one of the most efficient transformers, yet of the energy put into the production of a note by a vocalist only about 1 per cent. goes to charm the audience and the remaining 99 per cent. is, from this point of view, completely lost.

The power actually radiated as sound by various musical instruments has been measured, and is given in the following table. The figures quoted are for maximum loudness unless otherwise stated.

Source.	Power in watts.
Orchestra of seventy-five performers . .	70
Bass drum	25
Pipe organ	13
Snare drum	12
Cymbals	10
Trombone	6
Piano	0·4
Bass saxophone	0·3
Bass tuba	0·2
Double bass	0·16
Orchestra of seventy-five at average loudness .	0·09
Piccolo	0·08
Flute	0·06
Clarinet	0·05
French horn	0·05
Triangle	0·05
Bass voice	0·03
Alto voice *pp*.	0·001
Average speech	0·000024
Violin at softest passage	0·0000038

The first thing that impresses us in this list is the enormous range of power from the full orchestra playing *ff* to the solo violin playing *pp*. The ratio is about twenty million to one. It is interesting to note also how much of the power of the orchestra is radiated from one instrument—the bass drum. It supplies more than a third of the total power radiated. Ordinary conversation is carried on with very little power. It would take the power radiated by two million people in conversation to run a 50-watt electric bulb.

Power and Intensity.—If we assume that a source of sound of known power is placed in mid-air and radiates sound equally in all directions, it is easy to calculate approximately the intensity at any given distance. We define the intensity of waves at a point as the energy crossing a sq. m. of area (normal to the direction of propagation at the point) per second or, what is the same thing, as the power transmitted across the same area. If the source has a power P, then at a distance r this power is being transmitted across the surface of a sphere of radius r, and therefore of surface area $4\pi r^2$, and the intensity is $\dfrac{P}{4\pi r^2}$. The problem is, of course, not generally so simple as this. Sources are not usually suspended in mid-air, they do not radiate uniformly in all directions, and some energy is dissipated in the passage of sound-waves through the air, and disappears as sound. Nevertheless the example illustrates the relation between power and intensity and shows that, other things being equal, the intensity varies directly as the power of the source and inversely as the square of the distance from the source.

Intensity and Loudness.—It is found that, as we should expect, the greater the intensity I, the louder the sound, so that intensity and loudness are closely related. The power of the source depends, of course, on the amplitude of its vibrations, so that for a given source the greater the amplitude of vibration the greater the power, the greater the intensity in the neighbourhood of the source, and the greater the loudness. The intensity at any point is proportional to the square of the pressure amplitude at that point, and it is sometimes convenient to use pressure amplitude as a measure of intensity. The pressure amplitude $\hat{P}$ is the maximum excess of pressure over normal as the sound-wave passes, and the ' root mean square ' value of the amplitude is sometimes used for reasons which do not concern us here. It is denoted by P and $P = \dfrac{\hat{P}}{\sqrt{2}}$.

If we take a telephone diaphragm and set it in vibration by passing an alternating electric current through its coils, we shall find that for very small values of the current no sound is heard. As we increase the current the sound becomes audible, and we say we have reached the ' threshold of audibility '. The intensity required to reach this threshold varies with the frequency, as can be seen from Fig. 3.1. It is least at about 3500–4000, and from this frequency it increases in both directions. At the frequency corresponding to maximum sensitiveness the just audible intensity is exceedingly small.

When we come to consider the relationship between intensity and loudness, we can see at once that the one can hardly be simply proportional to the other. Since under given conditions the intensity is proportional to the power of the source, then (see p. 34) the intensity due to a full orchestra in a loud passage must be about twenty million times as great as that of the solo violin in a very quiet passage. Now, there is a great difference in loudness; but twenty million times as loud? Surely not! Here there comes to our rescue a law which is more or less true for all sensations—pressure, sight, &c.—that the intensity of the sensation is proportional not to the stimulus but to the logarithm of the stimulus. This means that every time the stimulus is multiplied by the same factor the sensation goes up the same number of steps. Thus, if the stimulus changes successively from 10 to 100, 100 to 1000, and 1000 to 10,000, &c., the sensation goes up by three equal steps. We get equal increments of loudness, not by adding equal increments of intensity, but by multiplying intensity always by the same factor.

There is, as we have seen, a limiting value for the intensity below which no sound is heard at all. Let us call this threshold value I_0. This will correspond to the zero of our scale. The sensation level of a note on this scale may be defined by the equation

$$S.L. = 10 \log \frac{I}{I_0}$$

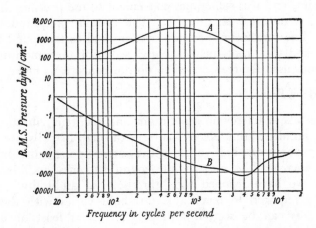

FIG. 3.1.—Limits of audibility for normal ears. *A*. Threshold of feeling (after Wegel). *B*. Threshold of audibility (after Fletcher and Munson)

where S.L. stands for sensation level, I for the intensity of the note, and 10 is chosen as a convenient multiplier to give the right size of unit on the scale. The unit is called the decibel. Since the logarithm of 2 is about 0·3, it follows that when the threshold intensity is double (I = $2I_0$), the sensation level goes up 3 decibels. This happens every time

PLATE III

From D. C. Miller's *Science of Musical Sounds* (Macmillan, N.Y.)

FIG. 2.3.—Helmholtz Resonators. These are tuned to different notes and were used for analysis

PLATE IV

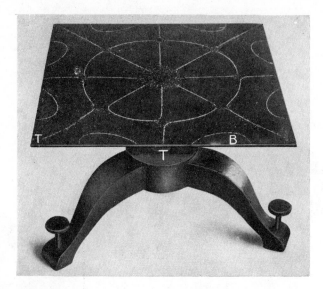

FIG. 5.2.—Sand pattern obtained with Chladni's plate (photograph)

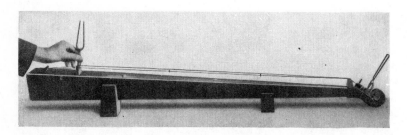

FIG. 5.3.—Monochord used to demonstrate partial tones

the intensity doubles, so that if the intensity is increased 2^6 or 64 times, the sensation level only increases by $6 \times 3 = 18$ decibels.

Strictly speaking, the threshold intensity I_0 varies considerably from person to person, and for the same person varies with age. It is necessary, therefore, to agree on a value to be used as the zero of the scale. The value chosen by international agreement is an intensity of 10^{-12} watt per sq. m., which corresponds to a R.M.S. pressure amplitude of 0·00002 Pa or Newtons per sq. m.

Can we now accept this scale as suitable for measuring loudness? Unfortunately the answer is no. The threshold of audibility is shown in Fig. 3.1, and will be seen to vary with the frequency of the note, and even if we allow for this, two notes which have the same sensation level measured for each above its own threshold of audibility have not the same loudness. Variations in the quality or timbre of notes also

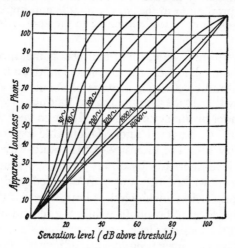

FIG. 3.2.—Relation between sensation level (intensity above threshold) and apparent loudness for various frequencies

cause slight differences in loudness for notes of the same pitch and intensity. These difficulties are overcome by choosing a note of definite frequency as a standard and measuring all other notes in terms of the loudness of this standard. The frequency chosen is 1,000 Hz, and when the loudness of any note is the same as the loudness of the standard tone at a definite sensation level in decibels, then this sensation level is taken as the loudness level of the note in question in terms of a new unit, the phon. Thus, if we wish to measure the loudness of any note, we take a source of frequency 1,000, whose loudness can be adjusted. Its loudness is then raised until it sounds equally loud with the note whose loudness is to be measured. The intensity level produced by the standard tone is then found in decibels, and this gives the loudness level of the note to be tested in phons.

Fig. 3.2 shows the relation between sensation level in decibels and loudness level on this scale in phons for notes of various frequencies.

Naturally we find that for frequency 1,000 the sensation level is always the same as the loudness level. That follows from our definition. But we notice that for frequencies of 500 and 10,000 this is nearly so. It follows that over this range the relation between intensity and loudness level is nearly the same as for the standard—i.e., the loudness level of a note goes up three phons when the sensation level increases by 3 decibels—i.e., when the intensity is doubled.

For the lower frequencies loudness level is not proportional to sensation level, and if we examine the curve for frequency 50 we shall see that when the loudness level in phons is 20, 40, 60, 80, 100 respectively, the sensation level in decibels is 12, 20, 25, 33, 48 respectively. Thus up to a sensation level of 33 dB. the loudness level rises much more rapidly than the sensation level. It follows that if a complex source of sound has its intensity increased, the increase in loudness level of the bass is out of proportion to that of the middle and upper frequencies. This fact may be noticed while listening to the approach of a band.

On the logarithmic scale, then, loudness is measured in a unit called the phon. Between the feeblest sound audible and the loudest sound audible there is a range of about 130 phons. The phon is about the smallest increment of loudness level which can be noticed under ordinary conditions, although very much smaller increments can be detected under favourable experimental conditions. Every time the intensity of the sound-waves is doubled the loudness level increases by about 3 phons. This happens, of course, if the power of the source is doubled and the circumstances are otherwise unchanged. Thus one twin crying may give a loudness level of 60 phons, while the second one joining in will only send the loudness level up to 63 phons—a quite negligible increase. If one lecturer produces a loudness level of 55 phons in his lecture-room it would take ten lecturers to send the loudness level up to 65 phons and 100 lecturers to send it up to 75 phons!

There seems to be little hope that the phon scale will be used to replace the indefinite *p* and *f* marking on a music score. Stokovski suggested the following equivalents:

ppp	20 phons		*f*	75 phons
pp	40 ,,		*ff*	85 ,,
p	55 ,,		*fff*	95 ,,
mf	65 ,,			

but his ideas have gained no general acceptance.

Sensation of Scale of Loudness.—A quite different method of constructing a scale of loudness is, of course, possible. It has been objected that a sound of 60 phons is not twice as loud as a sound of 30 phons. The physicist is inclined to reply to the critic with the question, ' Do you know what you mean when you speak of one note being twice as loud as another? ' The critic has been tested, and has stood the test unexpectedly well. Set to adjust one note to be half as loud as another, he has shown considerable consistency in his adjustment, and various observers show a large measure of agreement. Here, then, is the possibility of a scale based directly on judgements of loudness. This sensation scale is known as the sone scale of loudness as opposed to the phon scale of loudness level, but is not in wide use.

Noise.—Noise is a term which used to be applied very much in the sense of unmusical as we have defined it in Chapter I. It was reserved for sounds that were unpleasant, rough, and lacking definite pitch. It has now been defined by international agreement as ' sound which is undesired by the recipient '. This is a distressingly subjective definition, but, after all, we are dealing with a subjective term. The sound of the siren is not a noise if it is blowing the ' all clear ', and a Bach Fugue on the piano next door, however faultlessly performed, *is* a noise if we are trying to settle ourselves to sleep.

It is obvious from the definition of loudness level that the phon scale is applicable to noises as well as to notes. We can adjust the standard 1,000 Hz tone to be as loud as any particular noise, read off its intensity level, and so find the loudness level of the noise in phons. As a matter of fact there are now meters on the market which attempt to read the loudness on a phon scale directly and make it possible to follow the changes. Some meters even record an approximate phon value on a strip of paper which is valuable for rapidly changing sounds.

The measurement of noise is the first step towards its control, and this step is now being taken. There is perhaps insufficient evidence available to enable us to assess accurately the effects of noise on the health of the people. Two facts, however, are already clear: (1) that noise is in some degree detrimental to health, (2) that much of the noise to which we are ordinarily exposed is unnecessary. Before the beginning of the war there was an active movement on foot, sponsored by distinguished physicians and others, for the suppression of unnecessary noise. This involves suppression of noise at the source as its first line of attack. The road-drill, the aeroplane engine, the car, wheel traffic, the tubes, the milk-cart, the neighbour's radio—all these sources of noise were studied, improvements effected, and legislative control in some cases introduced. Protection against the passage of noise from the outside of a building to the inside and from room to room inside the building is the second line of attack, and this involves care in the siting and design of the building and special precautions in the material and details of its construction. The protection of the concert-room against noise from outside and the insulation of music practice-rooms from one another are our special interests in this field.

The following table gives some idea of the average loudness level of various noises:

	Phons.
Aeroplane engine 3 m away	130–140
Riveting machine 10 m away	102
Pneumatic drill 3 m away	90–100
Passing train a few metres away . . .	90–95
Traffic in front of St. Paul's	85–90
Traffic in Fleet Street	80–85
Conversation	60
Very quiet suburban street	40
Quiet whisper	20

Circulation.—It is a matter of common observation that the sound given out by an unmounted vibrating tuning-fork is ordinarily very small, but that if the shaft is pressed on to a table or if the fork is mounted on a box the sound is very greatly increased. The feebleness of the sound in the first case is due to ' circulation '. Loudness, as we have just seen, depends on sound intensity, and this in turn depends on pressure amplitude. Now, it is possible for air to vibrate in such a way as almost completely to avoid changes of pressure. In the case of a fork with fairly thin prongs, for instance, the air may move to and fro round the prongs, circulating without being compressed and rarefied. Obviously if a layer of air remains in front of a prong it will get compressed between the prong and the further layers until it is able to relieve the compression by expanding into the layers in front. On the other hand, as the prong moves forward it tends to leave a rarefaction behind, and if the air in front can move round into this, very little change of pressure will take place either in front or behind. Fig. 3.3 shows the path of the air as the prongs vibrate. If now a card be placed in the position shown, it will prevent circulation on one side, and the sound of the prong will be noticeably strengthened.

In the case of the diaphragm of a loud-speaker it is sometimes necessary to surround it with a 'baffle-board' several feet in diameter to prevent circulation in the case of low notes. The baffle-board plays the part of the card in the experiment on the tuning-fork.

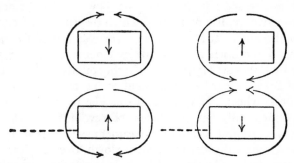

FIG. 3.3.—Circulating currents of air round the prongs of a vibrating tuning-fork. Insertion of card in position shown by dotted line stops part of circulation and increases loudness

The same thing is even more true of a stretched string or wire. If it is stretched between quite rigid supports almost no audible sound is produced when it is plucked. The sound produced by bowing a violin string comes not from the string, but from the body of the violin. Efficient radiation requires a source of large area, and the slower the vibrations, and therefore the lower the note, the greater must the area be. It will be noticed, for instance, that a tuning-fork radiates much more effectively the high, ringing note given out when it is struck on a hard surface than the much lower note which is its own fundamental tone. Thus all musical instruments combine a generator, which determines the frequency of the note and its loudness, with a radiator designed to prevent circulation and give increased loudness. The rôle of radiator is played as follows:

Generator.	Radiator.
Tuning-fork	Resonance box
Piano strings	Sound-board
Violin or 'cello strings	Body of instrument
Clarinet reed	Body tube
Vocal cords	Mouth cavities

CHAPTER 4

PITCH

Pitch and Frequency.—Pitch is the characteristic of a sound in virtue of which we describe it as ' high ' or ' low '. More exactly it is ' that subjective quality of a sound which determines its position in the musical scale '. It has long been correlated with frequency of vibration. Experiments were made by Robert Hooke (1635–1703) and later by F. Savart (1791–1841) on the sounds made by allowing a strip of brass or card to press against the teeth of a revolving, toothed brass wheel. It was found that the pitch of the resulting note depended only on the speed of revolution of the wheel. Anyone can carry out a similar experiment with a postcard and a meat-saw. When the card is drawn slowly over the teeth every separate tap is heard; but when it is drawn more rapidly the separate taps combine into a note, and the more rapid the movement of the card the higher the pitch of the note. The same fact can be established by the use of some form of siren (see next section). But a further fact can also be established—that whether the sound is produced by a toothed wheel or by a siren, and whether the siren is driven by air or by water, the pitch is always the same when the frequency is the same. This statement is almost universally true, although exceptions have been noted, and will be discussed later.

The lowest frequencies used in the orchestra are those of the double bass and the bass tuba, which lie between 30 and 40 Hz, while the highest frequencies occur as partial tones, and lie in the neighbourhood of 10,000 to 15,000 Hz. Tones of male speech embrace a range of from 120 to 8,000 Hz, those of female speech from 200 to 10,000 Hz.

The following table shows the notation generally adopted for the indication of pitch and the corresponding frequencies. These frequencies are given for the tempered scale as now used with a' at a frequency of 440 Hz. The frequencies are correct to the first decimal place.

From C to c is the 8-ft. octave denoted by CDEFGAB. From c to c' is the 4-ft. octave denoted by *cdefgab*, and so on. The 16-ft. octave (below C) is indicated by $C_1D_1E_1F_1G_1A_1B_1$. The designation of the octave as 4 ft., 8 ft., 16 ft., &c., has reference to the approximate length of the organ pipe which gives the C for that octave.

Staff	.	.				
Helmholtz	.	.	C	c	c′	c″	c‴
American	.	.	C_2	C_3	C_4	C_5	C_6
Frequency	.	.	65·4	130·8	261·6	523·3	1,046·5

The frequencies of the other notes of the tempered scale in the octave c′–c″ are as follows to seven significant figures:

c′	.	.	.	261·6256	g′ . . .	391·9954
d′	.	.	.	293·6648	a′ . . .	440·0000
e′	.	.	.	329·6276	b′ . . .	493·8833
f′	.	.	.	349·2282	c″ . . .	523·2511

The notes in the other octaves can be got by multiplying these numbers by 2 or dividing them by 2 the appropriate number of times.

The *ut* of the French system of pitch nomenclature is the first word of a Latin Hymn, Fig. 4.1, probably written about the year 770 A.D. The symbols of the next five notes of the octave are the first syllables in successive half lines—re, mi, fa, sol, la—and the next, si, is made from the two initial letters of Sancte Ioannes, the name with which the hymn ends. The idea is due to a Benedictine monk, Guido of Arezzo, who lived in the eleventh century. Ut, re, mi, fa, sol, la all fall on the proper notes in the scale. The si was added later. In Italy ut was replaced by Do—the initial letters of *dominus*—and in this form it is used in this country as the ' tonic sol fa ' system.

From time to time it has been noticed that pitch is not simply dependent on frequency, but depends also on loudness, and

FIG. 4.1.—Origin of the tonic sol fa system.

instances have been given of the lowering of pitch with increased loudness even when the frequency remains the same. An investigation described by Stevens [1] was carried out on three observers, and it was found that all agreed qualitatively in their judgements, although they differed somewhat in their quantitative estimates. In each case they were presented with a standard tone and with another tone of nearly the same frequency, and told to adjust the loudness of this second tone until its pitch was judged to be the same as that of the standard tone. It was found that the pitch of high tones rises with increased loudness, while the pitch of low tones falls with increased loudness. The division comes about the frequency 3,500 Hz, which is the frequency of greatest sensitiveness for the ear. Later observations by Snow [2] confirm these results generally for the lower-pitched tones, but emphasize the individual differences. Three observers out of nine failed to detect any change of pitch with loudness, but three others experienced changes greater than 35 per cent. at the greatest loudness used—a pitch change equivalent to an interval of about $2\frac{1}{2}$ tones. In all these cases pure tones were used. Experiments with the notes of the violin and violoncello show no consistent results, and it may well be that the greater complexity of the note explains the difference observed.[3]

Determination of Frequency.—The determination of the frequency of vibration corresponding to a sound of given pitch seems first to have been undertaken by Mersenne (1588–1648). He experimented with a hemp rope over 30 m long and with a brass wire 43 m long. In these experiments the vibrations were so slow that he could count them by eye. He varied the length and the tension, and ascertained how the frequency of the vibrations varied in consequence. In this way he obtained the laws of the transverse vibrations of strings (see p. 90) which can be summed up in the formula

$$f = \frac{1}{2l}\sqrt{\frac{P}{m}}$$

where　　　f = frequency,
　　　　　l = vibrating length in m,
　　　　　P = stretching force in Newtons,
　　　　　m = mass of 1 m of the string or wire in kilograms.

[1] *Journ. Acous. Soc. Amer.*, Vol. 6, p. 150 (1935).
[2] *Ibid.*, Vol. 8, p. 14 (1935).
[3] *Foundations of Modern Auditory Research*, Vol. I, J. V. Tobias, Ed., p. 424. Academic Press 1970.

He tuned a short brass wire to unison with one of the pipes of his organ, and, applying the laws he had discovered, he calculated the frequency of the wire to be 150 Hz.

Exactly the same method is available today. A sonometer is used which consists of a wooden base on which can be stretched a string or wire fixed at one end. The wire passes over a movable bridge, and its free end is attached to a scale-pan in which weights can be placed. The base may be set vertically, so that the weight exerts a direct pull, or horizontally, in which case the wire must pass over a pulley. A suitable wire and suitable tension having been selected, the vibrating length is varied by the movable bridge until the wire is in unison with the source of sound the frequency of which is to be measured. The known values for P and m are then inserted in the formula and f calculated.

Another method is based on the siren. The type of siren used consists essentially of a rotating disc in which is perforated an equidistant series of holes. If air is directed on to this row of holes from a jet, the escape of the air is periodically interrupted and a note is heard. The frequency of interruption of the jet, which is, of course, the frequency of the note, can be determined if we know the number of holes and the rate of revolution of the disc.

For laboratories plentifully supplied with electronic equipment, there are many further methods of measuring frequency, some of them of very considerable accuracy. Indeed frequency is one of the few physical quantities that can be measured to accuracies of the order of 1 part in 10^9 with relative ease. The simplest electronic way to compare the frequency to be measured with that from an accurate electronic oscillator is to convert the sound energy into electrical energy with a microphone and use that energy after amplification to deflect the electron beam of a cathode-ray oscilloscope in a vertical direction. At the same time, the output from the oscillator is used to deflect the beam horizontally. This arrangement leads to the formation on the screen of the patterns known as Lissajous' Figures, of which the simplest are the ellipse (in various degrees of eccentricity depending on the phase relationship between the two frequencies) if the frequencies are equal, and a figure of eight if the frequencies are in the ratio 2 : 1. If the frequencies are not in exact ratio, the pattern appears to revolve slowly; by observing the slow movement frequencies may be measured to an accuracy of about 0·1 Hz, assuming that the measuring oscillator is very accurate.

The frequencies of an oscillator and the unknown may also be compared by the method of beats (see p. 20); if the two electrical signals are mixed together in a multiplying circuit beats are found which may be observed on an oscilloscope or similar measuring instrument.

Another device which indirectly compares the frequency of the unknown with that of a standard oscillator is the frequency counter. In this an accurate standard of frequency (usually a quartz crystal oscillator) is used to create definite intervals of time, for example 0·1, 1, and 10 secs. The unknown frequency is fed into a ' gate ' circuit which is ' open ' only during the accurately measured interval of time. The resulting signal is then fed to a counting circuit which counts the number of periods in it. Thus the frequency of the original unknown signal is directly measured and may be displayed on meters or special display tubes. If the frequency to be measured is steady enough, frequency counters can be used to measure to an accuracy as great as a few parts in 10^8, which is far greater than that needed for musical purposes. Care has to be taken with complex waveforms (e.g., P on Fig. 5.7) to ensure that the fundamental frequency and not one of the harmonics is being counted.

Other methods using electronic equipment rely on comparing the period of the unknown with the time constant of a circuit composed of resistance and capacitance. Instead of using an oscillator to provide the horizontal deflection on a cathode ray oscilloscope as described above, a ' sawtooth ' waveform with which the beam is swept across the screen at a constant velocity determined by the time constants of the ' sweep circuit ' (usually included in the oscilloscope), and then rapidly returns and sweeps again, may be used. In this way a graphic representation of the pressure or velocity disturbance (depending on the type of microphone employed) against time is obtained, similar to that of the phonodeik (p. 69). Measuring the frequency then becomes a simple matter of measuring the length of one period of the waveform as displayed on the oscilloscope screen.

An even simpler device for waveforms that are not too complex is the frequency meter. In this device the signal from the microphone is converted to a waveform of standard amplitude and shape, and is then rectified so that only the positive going part of the waveform is retained. If this signal is then used to charge up a capacitance in a circuit of known time constant, the voltage across the capacitance is proportional to the input frequency, which may then be read on a normal voltmeter suitably calibrated.

Pitch Standards.—There are many reasons which suggest the desirability of fixing a standard pitch—i.e., assigning to some note of the scale a definite frequency.   For one thing, the musical effect of a performance depends on the pitch at which it is played.   The human voice is particularly sensitive in this respect, and if the pitch demanded for a vocal piece is uncomfortably high, more effort is involved and the quality is less satisfactory.   But, whatever instrument is used, the musical effect varies with the pitch to which the instrument is tuned, and the effect designed by the composer is achieved only if the pitch of the instrument is that for which the music was written.   Moreover, if instruments are to play together they must be tuned to an agreed pitch, and it is obviously desirable that this should always be the same, so as to avoid large variations in pitch when tuning.   In view of these and other considerations it is rather remarkable that for centuries there was no general agreement as to a standard of pitch, even nationally, let alone internationally.   The following table sets out the frequency corresponding to the A of the treble clef—what we have called a'— in some well-established cases, the numbers being taken from the *History of Musical Pitch* by Alexander Ellis, an abstract of which is given by Ellis as Appendix XX, Section H, in the *Sensations of Tone*, and tabulates nearly 300 observations.

	Date.	Frequency of a'.
Halberstadt organ .	1361	505·8
Church pitch, Heidelberg .	1511	377
„ „ North Germany	1619	567·3
„ „ Paris	1648	373·7
Schnitger's Organ, Hamburg .	1688	489
Paris Opera .	1699	404
Silbermann's Organ, Strassburg	1713	393
Handel's tuning-fork	1751	422·5
Bernhardt Schmidt's Organ, Cambridge	1759	395·2
Paris Opera .	1810	423
London Philharmonic Orchestra	1826	433
Paris Opera .	1858	428
French standard pitch (diapason normal) .	1859	435
Covent Garden Opera	1879	450
Philharmonic Society	1896	439
Piano manufacturers	1899	439
Military bands (Army Council)	1927	439

Standards of pitch must, in the first instance, have been fixed by assigning to an organ-pipe of some standard length a definite note on the scale, or even, to avoid transposition of a composition to some more convenient key, two alternative notes. Thus Schlick (1511) gives the length of a pipe to which either the note c or the note F might be assigned. Although the width of the pipe and the blowing pressure are not given, Ellis estimates that the former would correspond to a' 504·2 and the latter to a' 377, thus yielding a high pitch and a low pitch of which numerous contemporary and later examples can be quoted. Solo instruments combined in bands occasionally played with the organ and occasionally in the ' chambers ' of their patrons. The variety of pitch caused trouble, and Praetorius (1571–1621) suggested a ' suitable pitch ' of a' 424·2 in 1619. This pitch—sometimes called the Mean Pitch—agrees with Handel's own fork (a' 422·5 in 1751) and the London Philharmonic fork (a' 423·3 in 1820). This pitch prevailed for about two centuries—the period of Handel (1685–1759), Haydn (1732–1809), Mozart (1756–1791), and Beethoven (1770–1827). It is the pitch for which their compositions were written. Early in the nineteenth century, however, a rise in the standard of pitch began. It was due to the development of the brass instruments of the military band. These were found to give more brilliant effects at higher pitch, and the standard pitch rose to a' 448 at the Paris Opera in 1858 and to a' 456·1 at Vienna.

In 1859 a Commission was appointed by the French Government which included Berlioz, Meyerbeer, and Rossini amongst its members, and which selected a' 435. This selection was embodied by Lissajous in a standard fork ' diapason normal ', subsequently determined to be a' 435·4. This is the only legal standard in existence. A similar standard had been adopted in England some time before, the London Philharmonic Society about 1820 having adopted a' 433 on the recommendation of Sir George Smart. This was the so-called compromise pitch between the ' mean pitch ' or ' suitable pitch ' of the seventeenth and eighteenth centuries and the excessively high pitch which had succeeded it.

Once more, however, pitch started to rise, the pressure again coming from the military bands. In 1896 the Philharmonic Society adopted a' 439, and this standard was endorsed by the pianoforte trade. In 1927 the Army surrendered and lowered its standard so that general agreement became more possible.

The development of broadcasting stimulated and confirmed what was in any case a growing desire for a new agreement—and if possible an international agreement—on a standard of pitch. Measurements carried out by the Physikalische Technische Bundesanstalt on the pitch of broadcast performances in various countries emphasized the need. Performances were found to range between a' 430 and a' 460. The mean frequency for a' was highest in England (443·5) and in Czechoslovakia (443) and lowest in Denmark (439·5) and Holland (439·9). Variations during a performance have also been measured and recorded. In a performance of Beethoven's Piano Concerto No. 1 the pianoforte was at a' 440. The orchestra tuned to this and accompanied the pianoforte at this pitch, but in its tutti passages went up to 442. In the case of a performance of Bach's Oboe Concerto the solo instrument and orchestra tuned to 440, but the piano came in at 435. The oboe dropped to 439, at which level the orchestra accompanied it in its solo, dropping to 435 to accommodate the piano when necessary.

In May 1939 an International Conference was held in London. It was perhaps the last successful international effort before the outbreak of war. France, Germany, Great Britain, Holland, and Italy were represented, and memoranda were sent in by Switzerland and by the United States of America. The conference unanimously adopted ' 440 cycles per second for the note A in the treble clef '.

Pitch and Temperature.—The pitch of most musical instruments varies with temperature, and this variation is notable in the case of the wind instruments of the flute type. In this case it is due to the variation of the velocity of sound with temperature, a higher temperature causing greater velocity and therefore a higher frequency for the same wave-length, the wave-length being determined by the dimensions of the instrument (see Chapter 8). Thus a flute must be constructed so as to give a' 440 at a definite temperature. The French Commission in 1859 chose 59° Fahrenheit as the standard temperature. The temperature recommended by the British Standard Institution is 20° C. or 68° F. For the flue-pipes of organs and instruments of the flute type the frequency rises by about 1 part in 500 for 1° C. rise in temperature. The new international standard pitch of a' 440 at 20° C. would, for these instruments, correspond very closely to a' 435 at 15° C., which is the diapason normal. Reed-pipes of the organ and orchestral wind instruments are subject to about half this rise.[1] The pitch of pianos varies very little with temperature, the frequency falling by about 1 part in 10,000 for 1° C. rise in temperature.

It is this dependence of frequency on temperature which is sometimes responsible for a steady rise in pitch as an orchestra gets warmed up. The rise is produced not only by the change in temperature, but also by the increase of the moisture content of the air. It is probable that in the future air-conditioning may be more widespread and changes in temperature and moisture content less marked and therefore less troublesome, although it must be remembered that it is the player's breath which is responsible for the warming-up process, and this must always cause initial changes.

These considerations have to be borne in mind when selecting an instrument which can embody the standard frequency and be relied upon to give it when the circumstances are suitably arranged. In most ways the best instrument for the purpose is a tuning-fork. Its temperature coefficient is very small, the drop in pitch for a steel fork being about 1 part in 9000 for 1° C. rise in temperature. Forks made of elinvar have a negligible coefficient and do not rust. Pitch-pipes were at one time in common use, but have a much larger temperature coefficient. The oboe, although very often used when an orchestra is tuning up, is most unsuitable. Electronic generators may be substituted for the oboe, but they do not seem to be very popular among musicians. A number of broadcasting authorities transmit the internationally agreed standard frequency before their daily programmes, which may be used to calibrate the more convenient, portable, tone generators actually in use in the concert hall.

Musical Intervals.—A note of any definite pitch being chosen as a starting point, a number of other notes can be picked out by the ear as in some way simply related to it. This kind of relationship was first studied by Pythagoras (572–497 B.C.) in the earliest acoustic experiments recorded. He showed that the simplest and most obvious of all such relationships—that between a note and its octave—is always obtained with the two segments of a stretched string when it is divided by a movable bridge so that the ratio of the lengths of the segments is 2 : 1. He found another familiar relationship between the notes given by the two segments when the ratio of the lengths was 3 : 2. Whatever the tension or thickness of

[1] See R. W. Young, *J. Acoust. Soc. Amer.*, Vol. 17, p. 187 (1946).

the string, the same recognizable relationship always holds between the notes given by two segments of the string. Their pitch
may vary, but the relationship between them remains the same.
Pythagoras found that whenever the string was divided by the
movable bridge so as to give two notes in *some* recognizable relationship, the ratio of the lengths of the segments was always the
ratio of two small whole numbers. Such notes are said to form a
musical interval, and the interval is usually named from the position
of the two notes in the musical scale. Thus, if we take the scale
of C major and number the notes from the keynote, *c*, then we have
the white notes of the pianoforte:

c	*g* fifth
d second	*a* major sixth
e major third	*b* major seventh
f fourth	*c'* octave

For the minor scale with *c* as keynote we can take

c	*g* fifth
d second	*a♭* minor sixth
e♭ minor third	*b♭* minor seventh
f fourth	*c'* octave

The fourth and fifth are common to both scales. The second is
also common to both scales, but hardly recognizable as a simple
musical interval. Since the third, sixth, and seventh are different
on the two scales, we distinguish them as major and minor. The
thirds, the fourth, the fifth, and the sixths are all more or less recognizable musical intervals, even to people without musical training.
They all satisfy the law of Pythagoras that they are produced by
two segments of a stretched string when their lengths are in a simple
numerical ratio. He regarded this fact as a sufficient reason for
the relationship, the pleasing effect being due to the simplicity of
the ratio. Later experiments were made with a siren perforated
with four rows of holes, the numbers of the holes in successive rows
being in the ratio of, say, 4, 5, 6, and 8. If the wind can be directed
to any row at will, then it will be found that the first and last rows
always give the interval of the octave, whatever the speed of revolution of the perforated disc. If the speed is increased, both notes
rise in pitch, but the interval remains an octave. The first and
third rows give the interval of the fifth; the third and fourth rows
give the interval of the fourth; the first and second rows give the
major third; the second and third rows give the minor third.

Thus, whatever the absolute pitch of the notes concerned, two notes will always appear to the ear to be in some simple relationship if their frequencies are in a simple numerical ratio. This relationship is apparent when the notes are sounded successively.

Thus, assuming the key note to be *c*, we have:

Notes.		Interval.	Frequency ratio.
c	*c′*	Octave	1 : 2
c	*g*	Fifth	2 : 3
c	*f*	Fourth	3 : 4
c	*e*	Major third	4 : 5
c	*e♭*	Minor third	5 : 6
c	*a*	Major sixth	3 : 5
c	*a♭*	Minor sixth	5 : 8

These intervals give us all the notes of the major diatonic scale except *b* and *d*. If we start our scale on *g*, then *b* is the third and *d* is the fifth.

Using these ratios and the standard pitch for *a′*, we are in a position to calculate the frequencies of all the notes in the ' natural ' or ' true ' scale of C major—i.e., the untempered scale of C. These are:

c′	.	.	.	.	264	*g′* .	.	.	.	396
d′	.	.	.	.	297	*a′* .	.	.	.	440
e′	.	.	.	.	330	*b′* .	.	.	.	495
f′	.	.	.	.	352	*c″* .	.	.	.	528

The difference between these frequencies and the frequencies of the tempered scale already given (p. 43) is discussed in Chapter 11.

Measurement of Pitch Intervals.—It is obviously desirable to have some scale on which we can measure the size of a musical interval in the sense of the range of pitch it embraces. We want to know the relative sizes of the octave, the fifth, the fourth, &c. Now, we have just seen that these are defined by the ratios 2 : 1, 3 : 2, 4 : 3 or 2·00, 1·50, and 1·33. These ratios give us the order of magnitude of the intervals, but not their relative size. And if we wish to add two intervals, we have to multiply the corresponding ratios. Thus, if we wish to add a fifth and a fourth, we see that the frequency of the upper note of the fifth bears to the frequency of the lower note the ratio 3/2. Now, the upper note of the fifth is the lower note of the fourth which is to be added, and the frequency ratio for the fourth is 4/3. The frequency of the upper note of the fourth must therefore bear to the frequency of the lower note of the fifth the ratio $4/3 \times 3/2 = 2$. It is, therefore, the octave—a result which we should not have anticipated if we had tried to add 1·50 and 1·33. We have the same difficulty in subtracting one interval from another. We have to divide the corresponding ratios. It is therefore very desirable to choose a unit in terms of which all musical intervals can be measured and the measurements added or subtracted in the ordinary way. Such a unit is the 'cent', which we shall define in such a way that there are 100 cents to the tempered semitone, 200 to the tempered tone, and 1,200 to the true octave. It is a convenient size of unit, as it represents approximately the smallest pitch interval that the very best ear can appreciate. In terms of the cent the intervals given above are as follows:

minor third . . 315·6	minor sixth . . 813·7
major third . . 386·3	major sixth . . 884·4
fourth . . . 498·0	octave . . . 1200
fifth . . . 702·0	

The cent is derived as follows. Since musical intervals depend on frequency ratios, and since we require to have logarithms of ratios if we are to add and subtract in the ordinary way, we can say at once

$$\text{Pitch Interval} = K \log_2 f_1/f_2$$

where f_1 and f_2 are the two frequencies and K is a constant which we can choose in such a way as to get an appropriate size of unit. Thus, remembering that for an octave

$$f_1/f_2 = 2 \text{ we have}$$
$$\text{Octave} = K \text{ units.}$$

and we have to choose K so as to get the most desirable number of units in the octave.

If we put K = 1,200, then we have a unit called the cent—used by the translator of Helmholtz' work.

This scale has the advantage that there are 100 cents to the tempered semitone and 200 to the tempered tone. It is the scale most widely used in the published literature on intervals, and is also used on instruments designed to measure musical intonation.

If we put K = 100, we have the centioctave, so called because we now have

$$\text{Octave} = 100 \text{ centioctaves.}$$

If we put K = 1,000, we have the millioctave.

These units give very convenient numbers for the octave, but none of the usual intervals give us simple numbers.

The unit suggested by Savart is obtained by setting K = 300; this gives 25 s. to the tempered semitone, 50 s. to the tempered tone, and is widely used in the French literature. In the earlier editions of this book Wood used the savart but the change to the cent has now been made to reflect the greater use of this unit in the English speaking world.

Pitch Range of Musical Instruments.—The pitch range of musical instruments can be regarded from two different points of view. We may think of it as the range of the scale which can be produced on the instruments or, taking account of the fact that each note consists of a fundamental tone which gives the pitch and a series of partial tones of greater frequency and higher pitch not usually separately perceived by the ear, we can look for the limits of frequency outside of which partial tones make no noticeable contribution to the quality of the tone of the instrument. It is clear that this second point of view is likely to give a very different result for the upper limit of the range. It is not at first sight clear that it can give a different result for the lower limit. But we must at least keep open the possibility that the fundamental in the case of the lowest notes may itself be relatively unimportant and the partial tones may be the only important contribution.

So far as the range of frequency for the fundamental tones is concerned, the instruments with the largest range are the piano and organ. The piano may go as low as 27·5 Hz and the organ as low as 16·5 Hz, but the musical character of these notes is doubtful and they are not used alone. The upper limit of the piano may be as high as 4224 Hz. The piccolo may use a note of 4752 Hz. Broadly we may say that for fundamentals the useful musical range is from 40 to 4,000, while the range of audibility is from about 25 to 20,000.

When we take account of the partial tones as well as the fundamentals, some very interesting results emerge. The method of test which has been used is as follows. The music of the instrument in question is reproduced electrically by a system which enables all frequencies below a certain limit, or all frequencies above a certain limit, to be filtered out. These limits can be varied. A team of expert musicians listens and decides at what limit of frequency the filter just begins to make a difference. For instance, the filter may be set to cut out all frequencies below 100. The observers are made to listen to the music, sometimes filtered and sometimes unfiltered, and asked to say in each case what the condition is. If they are right in 50 per cent. of their judgements, then the filter makes no difference—if they were guessing, they would be right to this extent. If they are right in 60 per cent. of their judgements, then the filter is making some difference, and if they are right in 80 per cent. of their judgements, the effect of the filter is marked. The figure (4·2) shows the results of the tests.[1] The musicians were instructed to play their instruments 'loud'. Tests were made with the instruments played in their several octave ranges or with their different techniques, to ensure satisfactory observations.

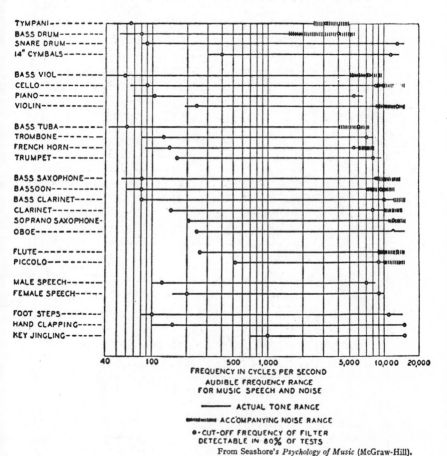

FREQUENCY IN CYCLES PER SECOND
AUDIBLE FREQUENCY RANGE
FOR MUSIC SPEECH AND NOISE

————— ACTUAL TONE RANGE

▬▬▬▬ ACCOMPANYING NOISE RANGE

●-CUT-OFF FREQUENCY OF FILTER
DETECTABLE IN 80% OF TESTS

From Seashore's *Psychology of Music* (McGraw-Hill).

FIG. 4.2.

[1] Snow, *Journ. Acous. Soc. Amer.*, Vol. 3, p. 161 (1931).

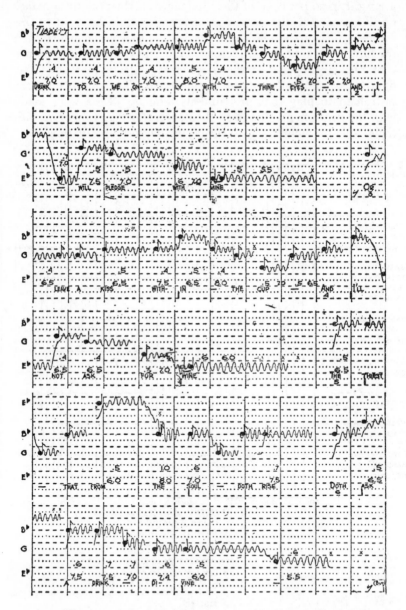

FIG. 4.3.—Vibrato analysis of 'Drink to me only with thine eyes'. Pitch is represented by a graph for each note on a semitone staff. The dots mark tenths of
seconds. The upper numbers give the average extent of pitch vibrato, the lower
number the number of pulsations per second.

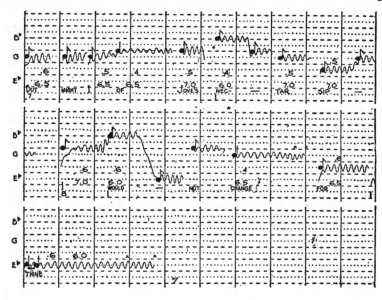

FIG. **4.3** (*continued*)

The piano was found to be the only instrument which did not require the reproduction of its lowest fundamentals for perfect fidelity. The lowest fundamental of the double bass (40 Hz) was required for perfect fidelity, but very little was lost in the case of any instrument so long as frequencies of 60 and over were transmitted. At the upper end of the range some attempt was made to distinguish between frequencies involved in the actual tone quality of the instruments and frequencies constituting the noises accompanying the tone—key clicks, lip noises, buzz of reeds, and hissing of air. As will be seen, many of the instruments produced noises that extended to high frequencies, but only the oboe, violin, and snare-drum were thought to extend their tone-ranges to the upper audible limit. An upper cut-off at 10,000 Hz did not affect the tone of most of the instruments to a marked extent but every instrument except the bass drum and tympani was affected when frequencies above 5,000 Hz were filtered out. For faithful reproduction the double bass required the greatest range (seven octaves) and the piccolo the smallest (four octaves).

Vibrato.—It is a well-known fact that many singers do not attempt to hold a sustained note at constant pitch, but produce a periodic variation of the pitch known as ' vibrato '. A good vibrato is said

to consist of ' a pulsation of pitch usually accompanied by syn-
chronous pulsations of loudness and timbre of such extent and rate
as to give a pleasing flexibility, tenderness, and richness to the tone '.
Vibrato is not confined to the human voice, it occurs also in string
tone.   So far as the voice is concerned, excessive vibrato is generally
condemned.   What is not perhaps fully realized is that vibrato is
present in practically every note of every song, whether the note is
long or short, high or low, weak or strong.   This is revealed by
records of pitch and loudness, as, for instance, that given in Fig. 4.3.
When the pitch variation in the vibrato is small it escapes observation as a variation in pitch at all, and is heard only as a specific
quality of tone.   The rate of variation is about 6·5 Hz, and the
range of the variation is about a semitone.   A variation in loudness—a loudness vibrato—is sometimes associated with the pitch
vibrato, but seems to be of secondary importance and subordinate
to the pitch vibrato.   The mean pitch coincides fairly accurately
with the true pitch.

The following figures are given by Seashore:

	Average rate per sec.	Average range of variation.
All artists (twenty-nine singers) .	6·6	0·48 tone
Galli-Curci	7·3	0·44
Caruso	7·1	0·47
Chaliapin	6·8	0·54
Tetrazzini	6·8	0·37
Gigli	6·5	0·57

In the case of the violin, the vibrato is present in practically all
tones produced by completely stopping the string.   There is, of
course, no pitch vibrato when the open string is used.   The frequency of the vibrato is about 6 Hz, and the extent of the variation
is about a quarter of a tone, and is fairly constant and regular.
Here again a loudness variation frequently accompanies the
pitch vibrato, but appears to be secondary.

It is important to emphasize the fact that vibrato in singing is
almost universal.   Recognized professional singers use it in 95 per
cent. of their notes.   Primitive peoples use it.   Even teachers
opposed to its use and professing not to use it show it when the
record of their performance is analysed.   The crudities of excessive vibrato should not be allowed to prejudice our judgement.[1]

[1] See also *Acustica*, Vol. 27, p. 203 (1972).

Absolute Pitch.—A sense of absolute pitch is the capacity to name a note which has been produced without any standard of pitch being previously sounded. It is not the correct judgement of a musical interval—it is assigning a note to its proper place in the musical scale. It has long been known that some people possess this gift, that those who do possess it differ very much in the accuracy of their judgements, and observations have been made suggesting that in some cases at least the accuracy of judgement can be considerably improved with practice. It is only recently, however,[1] that a systematic study of the subject has been made. Bachem made observations on 110 subjects reputed to be able to judge absolute pitch. The investigation brought out some very interesting differences, which were made the basis of a classification both suggestive and important. The subject of the test was made to stand with his back to the piano. A note was then sounded on the piano, and the subject was asked to turn round and point to the appropriate key. Tests of a different kind were also made with tuning-forks, violins, bells, whistles, &c. Of the subjects tested seven were found to have the gift in a quite remarkable degree. Their judgements were instantaneous, and they made no mistakes. They were equally true and certain in their judgements for all ranges of pitch and for all kinds of instruments. The judgement seemed to rest first on a characteristic quality or chroma attaching to the note, which at once fixed it as a C or a G or a B. The height of the pitch of the note then gave the proper octave.

Next to these came a group of forty-four subjects. These seemed to make their judgement in the same way, but were less accurate. Octave errors were common, no doubt due to getting the right note from its quality and assigning it to the wrong octave. Semitone errors were also common, and for each subject fairly systematic. Either they were all in one direction—suggesting that the subject had been accustomed to a different standard of pitch—or else they occurred near the ends of the range and the judgements were sharp at the lower end of the range and flat at the upper end. These subjects seemed to have the same gift as the first seven, and were able to make their judgements for most of the musical instruments tried. They were just less certain and accurate.

Next to these came a further group of thirty-nine, whose capacities showed marked limitations. Again their method of judgement

[1] Bachem, *Journ. Acous. Soc. Amer.*, Vol. 9, p. 146 (1937); Ward, *ibid.*, Vol. 26, p. 369 (1954).

seemed to be the same, and, with certain limitations, their accuracy was good. But their accurate judgements were either confined to certain instruments or to three or four octaves of pitch or in both ways.

An entirely different method of judgement marked the next group of thirteen. They seemed to have a standard pitch in their minds and to estimate the interval of the given nôte above this. The judgement was not immediate and certain, but slow and hesitating, and accompanied in some cases by humming. Bachem calls this ' quasi ' absolute pitch, and regards it as fundamentally different from the capacity shown by the ninety subjects previously discussed. This is borne out by the fact that the errors are different in kind as well as in degree.

Lastly comes a group the members of which make a judgement —usually a poor one—based simply on an estimation of tone height. These subjects cannot be said to have the sense of absolute pitch at all.

Further tests were made on a larger number of subjects [1] as a result of which it was concluded: (a) that there is good evidence in support of the view that the gift is hereditary, (b) that it cannot be acquired late in life and cannot be acquired at all unless a certain predisposition exists, (c) that it is more easily acquired by musicians than by others, and (d) that it occurs more frequently among the blind than among people of normal sight.

[1] Bachem, *Journ. Acous. Soc. Amer.*, Vol. 11, p. 434 (1940).

CHAPTER 5

MUSICAL QUALITY

Musical ' Quality '.—If a note of given pitch is played successively on two different musical instruments, and played with exactly the same loudness, we can distinguish between the two sounds and refer each to its appropriate instrument. The basis of this judgement is the ' quality ' of the sound. The piano, the violin, the voice, the flute, &c., each has its own characteristic ' quality '. The German word for it is ' klangfarbe ', the French word (frequently borrowed in English) is ' timbre '. But quality not only enables us to distinguish between two notes produced on different kinds of instruments, it also enables us to distinguish between two notes produced on two different instruments of the same kind. What distinguishes the voice of one singer from that of another and is characteristic of an individual voice is its quality. Or again, take a stretched string, pluck it near one end and listen to the note; then pluck it again, this time at the middle, and listen to the note; the pitch will be the same and the loudness can be made the same, but there will be a marked difference in quality.

To understand the cause of these differences of quality it is perhaps well to begin by reminding ourselves that sources of sound—e.g., stretched strings, air columns, &c.—have usually numerous possible modes of vibration. They can vibrate in a larger or smaller number of parts, and each mode of vibration has a different frequency. The greater the number of parts, the greater the frequency and the higher the pitch of the note.

One of the classical methods of experiment for demonstrating this is that of the sand figures on a vibrating plate devised by Chladni (1756–1827). He was a German physicist, and is said to have demonstrated these figures to Napoleon Bonaparte in a two-hour audience, as a result of which the Emperor gave him 6000 francs to enable him to prepare a French translation of his German treatise on acoustics published in 1809.[1] Chladni used a glass or metal plate as his source of sound. In one form the

---

[1] *Anecdotal History of Sound*, D. C. Miller, Macmillan, 1935.

plate is square and is fixed at the centre.   It is strewn with sand,
and is then set in vibration by applying a violin bow to one
point of the edge and touching another point.   The plate breaks
up into vibrating patches or segments separated by lines which
are at rest.   This is a particular case of the stationary vibration
referred to on p. 15.   The bow generates waves by alternately
sticking and slipping on the edge of the plate.   These waves
travel across the plate, are reflected from its edges, and become
superposed on the waves which are approaching the edges.   The
result is a pattern of nodal lines where the plate is at rest separat-

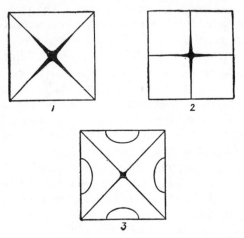

FIG. 5.1.—Sand patterns obtained with Chladni's plate
(diagrams)

ing segments of the plate which are in motion.   All the points
on the plate in one segment are vibrating together and are in
opposite phase to the segments on the other side of the nodal
line which separates them.   The sand gets thrown off the vibrating segments of the plate and gathers along the nodal lines.
Fig. 5.1 shows the three simplest modes for a square plate clamped
at the centre, giving the three lowest tones for the plate.   Fig. 5.2
is a photograph of one of the more complicated modes.   One
nodal line joins the centre to T, the point of the edge touched by
the finger, and the rest of the pattern develops by symmetry.
We find (a) that each pattern—i.e., each mode of vibration—
has its appropriate pitch, and (b) that the greater the number
of vibrating segments, the higher the pitch of the note.   These

notes—due to the vibrations of the plate in parts—are called its partial tones, and although by this method of experimenting they can be elicited one by one, when the plate is struck or dropped they are developed simultaneously, and the resulting sound is due to the mixture of partial tones.

If now we consider a stretched string, we find the same kind of phenomenon, but in some respects simplified. If a rubber cord, about 7 or 8 m long, is fixed at one end and the other end is held in the hand and moved gently to and fro sideways, it is easy to time the motion of the hand so as to set the rubber cord into vibration in its simplest mode. In this case the point of greatest amplitude of motion—the antinode—is at the centre, while the two ends are relatively still, and act as nodes. This is another case of stationary vibration. The simple to-and-fro vibration of the cord is the result of the superposition of waves communicated to the cord by the hand and reflected from the fixed end of the cord, upon the waves which are travelling from the hand to the fixed end but have not yet reached it. If now the hand be moved to and fro with twice the frequency, the cord breaks up into two vibrating segments with a node at the centre and two antinodes $\frac{1}{4}$ and $\frac{3}{4}$ of the length of the cord from either end. Moving the hand still more rapidly, so that now its frequency of motion is three times as great as before, we find the cord breaking up into three vibrating segments with nodes $\frac{1}{3}$ and $\frac{2}{3}$ of the length of the string from either end and antinodes at $\frac{1}{6}$, $\frac{1}{2}$, and $\frac{5}{6}$ from either end. This process can be extended while time and patience remain, and shows that the string may vibrate in 1, 2, 3, &c., segments and that the frequencies of these modes of vibration are in the ratio 1 : 2 : 3, &c. A series of notes the frequencies of which are in this ratio is called the harmonic series. The notes in the scale corresponding to the first six are the first, eighth (or octave), twelfth, fifteenth, seventeenth, and nineteenth, or, starting from c, they are: c', g', c'', e'', g''. The seventh partial does not occur in the scale. The eighth partial is the twenty-second note in the scale, c'''.

Of course the rubber cord does not give a musical note. Its most rapid vibration is too slow, and the hand is incapable of communicating a sufficiently high frequency. Let us instead take a thin wire stretched between bridges fixed to a wooden base, and attached to a peg by means of which the tension can be altered. A scale is attached so that fractions of the length of the wire can be measured. An apparatus of this kind (see Fig. 5.3) is called a monochord. The tension is adjusted until the frequency of vibration of the wire is, for example, 128. Three small paper riders are then placed on the wire at $\frac{1}{4}$, $\frac{1}{2}$, and $\frac{3}{4}$ of the length from one end. When a fork of frequency 128 is struck and its shaft held on one of the bridges, the wire is forced to vibrate in the frequency of the fork which is the fundamental mode for the wire, and all three riders are thrown off, showing that the wire was vibrating as a whole in one segment. If now the riders are replaced and the experiment repeated with a fork an octave higher (frequency 256), then the middle rider remains in position while the other two are thrown off, showing that the wire is now vibrating in its second mode—i.e., in two segments with a node at the centre, giving its second partial tone. A fork of frequency 384 will make it vibrate in three segments, and the position of the nodes can be shown by placing riders $\frac{1}{6}$, $\frac{1}{3}$, $\frac{1}{2}$, $\frac{2}{3}$, and $\frac{5}{6}$ of the length from one end. The first, third, and fifth riders will be thrown, and the second and fourth will remain in position. Forks of frequency 512, 640, &c., can also be used, but the point is demonstrated. The string has a series of possible modes of vibration. These modes involve the string vibrating in 1, 2, 3, 4, &c., segments, and the frequencies of the corresponding notes are in the ratio $1 : 2 : 3 : 4$, &c.

Analysis of a Musical Note by Ear.—When a string is plucked, or bowed, or struck, what is the nature of the resulting sound? Is it a single tone corresponding to the first mode of vibration of the string—the fundamental or first partial? Our ear suggests that it is. We are not generally conscious of hearing anything else. We have the impression of listening to a single tone. But a little careful attention and an attempt to analyse our sensation reveal the fact that we do not merely hear the one musical tone which dominates the sensation when received uncritically and determines the pitch of the note. We become aware of a series of constituent tones forming the harmonic series. There are numerous references in the literature of the subject to this phenomenon. Thus we find Aristotle (384–322 B.C.), Book XIX, Problem 8, asking, 'Why does the low note contain the sound of the high note?' And again in Problem 13, 'Why is it that in the octave, the concord of the upper note exists in the lower, but not vice versa?' Mersenne (1588–1648), a minorite friar, wrote the first extended treatise on sound and music, *Harmonie*

Universelle, published at Paris in 1636. In it he refers to these problems of Aristotle, and in a section headed ' To determine why a vibrating string gives several sounds simultaneously' (p. 208) he says: ' But it must be remarked that Aristotle did not know that the struck string gives at least five different sounds simultaneously, of which the first is the natural sound of the string, serving as fundamental to the others and to which alone attention is . . . paid in singing; all the more because the others are so weak that only the best ears can hear them easily. It is necessary to choose a deep silence in order to hear them, although this will no longer be necessary when the ear has become accustomed. As for myself, I have no difficulty; I have no doubt that anyone can hear them who gives the necessary attention. Now these sounds follow the ratio of the numbers 1, 2, 3, 4, 5 because four sounds are heard different from the fundamental, of which the first is the octave above, the second is the twelfth, the third is the fifteenth, and the fourth the major seventeenth.' That Mersenne associated these partial tones with quality is clear from a query in the same section whether ' the sound of each string is more harmonious and agreeable as it causes to be heard a greater number of different sounds simultaneously '. This is the first clear association of quality with partial tones.

We may perhaps pause here a moment to reflect on this capacity of the ear to analyse sounds. It is surely remarkable that when we listen to a vocal quartet, and the sound-waves from four sources bombard our ears simultaneously, we can, from the complicated resultant movement of our ear-drum, reconstruct the four separate sources of sound and follow any one of them at will. Even more remarkable, perhaps, is the capacity to isolate the solo instrument in a concerto when sixty or seventy sources of sound are simultaneously operating. But perhaps the most remarkable power of all is that which enables the ear to hear separately in the to-and-fro movement of the drum produced by a single note the partial tones corresponding to the vibrations of the source in one, two, three, &c., parts. How this analysis is accomplished will occupy our attention later (p. 89).

We can practise this power of analysis in a variety of ways. Using the piano, we can first demonstrate the existence of partial tones and then listen for them. If we depress the key *c'* without sounding the note, and then strike the key *c* and release it (keeping the key c' depressed), we shall hear the c' sounding out quite strongly. The string has been set in vibration by resonance with the second partial of c. If we depress successively the keys $b\flat$, b, $c'\sharp$, d' we shall find that in none of these cases does striking the key c give any resonance at all. The experiment can be repeated with c using the notes corresponding to the other partials—g', c'', e'', g''. Having accustomed the ear to the pitch of the partials and demonstrated their existence, the observer can practise direct analysis by first playing the note c' as a guide to the ear and then releasing the key and sounding c. With a little attention the c' can be heard as a constituent of the c, as also can g', c'', e'', and g'', although the hearing of the last two is more difficult.

Another experiment can be performed either with a mono-chord or with a violin, 'cello, &c. Pluck a string near one end and then touch it very lightly at its mid-point with the corner of a handkerchief. The octave (second partial) will sing out loudly. Damping has destroyed the fundamental, which requires a place of maximum motion at the centre, and so brings into prominence the previously unnoticed second partial which has a node at that point. If now the string is plucked as before, but this time *not* damped, and if we concentrate our attention on the octave, it will be quite clearly heard in the general mass of tone. The same thing applies to the third partial. Damping at one third of the length of the string from the end brings it into prominence. But when we have once heard it and fixed our attention on it, we can hear it clearly and distinctly in the ordinary note of the string.

Analysis by Resonators.—Another method of analysing a note into its components is by the use of resonators. This method is the same in principle as that given already for the piano, but the apparatus is different in design. The air contained in a cavity of any shape when set in vibration will give a note. The pitch of the note is raised by making the volume of the air less or the aperture connecting it with the outside air larger in area. The note given by a jug or bottle when it is being filled at a tap is due to the contained air, and there is an obvious rise in pitch as the filling proceeds and the volume of contained air gets less. The second point can be illustrated by taking a wide-mouthed jar and passing the hand to and fro over the mouth as it is being filled. Helmholtz designed a series of air resonators like that shown in Fig. 2.3. Each resonator is shaped somewhat like a turnip. The tapering end—which is open—is inserted into the ear, and the wide end is presented to the source of sound. If the natural tone of the resonator is present in the mass of tone which is received, it is reinforced and strengthened by the resonator and is heard by the ear. If, then, we take a series of resonators whose natural frequencies form the harmonic series for the note to be analysed, and use each of them in turn, we

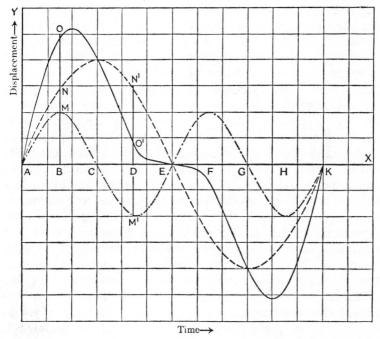

FIG. 5.4.—Composition of two displacement curves (dotted) representing a note and its octave, the lower note having twice the amplitude of the higher. The continuous curve represents the resultant displacement

shall ascertain which of these is present in the note, and may form some idea of the relative strength.

They may be used to assist analysis by the unaided ear if, while the note is sounding, the resonator is gradually withdrawn from the ear. We can still hear the partial tone sounding.

Analysis by Graph.—To every vibration there corresponds an appropriate graph or displacement curve (p. 5). Take a sheet of paper, make a pencil describe to-and-fro movements from side to side across it, and at the same time pull the paper down at right angles to the movement of the pencil. The point will trace a wavy curve on the paper which is the displacement curve or space–time graph for the motion of the pencil point, enabling us afterwards to find its displacement at any previous instant. If the pencil moves to and fro with the same kind of motion as a pendulum bob, the graph is the sine curve already studied— the graph corresponding to a ' simple harmonic vibration '. More complicated graphs can be built up by combining simple harmonic graphs. To do this we draw them to the same axis, add their ordinates at any point at which both curves are on the same side of the axis, and subtract them at any point at which they are on opposite sides. Fig. 5.4 shows the combined curve due to two simple harmonic curves. $BO = BM + BN$, $DO' = DN' - DM'$. It will be noted that one curve makes two complete vibrations, while the other makes one. The curves thus represent notes whose frequencies are in the ratio of $2:1$, i.e., they represent a note and its octave. The amplitudes or maximum displacements are also in the ratio of $2:1$, the lower note having twice the amplitude of the higher. Different resultant curves could have been obtained by taking the amplitudes in a different ratio or by shifting one curve along the axis relative to the other, which would be equivalent to altering the relative phase of the vibrations. By combining a larger number of constituents, we can, of course, get much more complicated vibration curves, even if we confine ourselves to the harmonic series—i.e., to curves corresponding to frequencies in the ratio of $1:2:3:4$, &c.

This leads us to the consideration of a very important mathematical theorem propounded by J. B. Fourier (1768–1830) in 1822. For our present purpose it may be stated as follows: any periodic vibration, however complicated, can be built up from a series of simple harmonic vibrations whose frequencies are in the ratio $1:2:3:4$, &c., and whose relative amplitudes and relative phases are suitably chosen. Conversely, any periodic vibration can be analysed into such a series, and the relative amplitudes and relative phases determined. Such a process applied to a complicated vibration is known as harmonic analysis. If the vibration to be analysed contains a good many components, the process may be a long and a difficult one. In order to facilitate this, much ingenuity has been devoted to the design of instruments which can carry out this analysis mechanically, and some of these ' harmonic analysers ' may be seen at

PLATE V

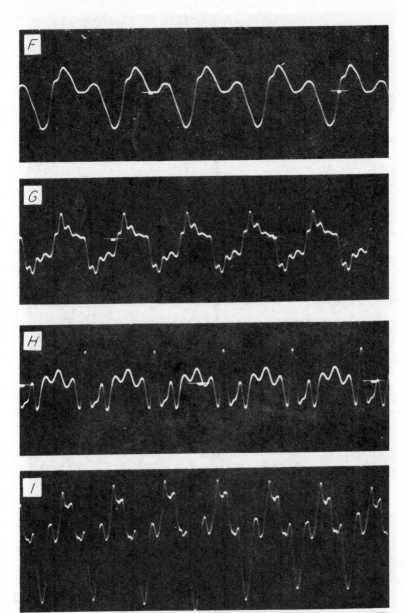

From D. C. Miller's *Sound Waves, Their Shape and Speed* (Macmillan, N.Y.

FIG. 5.6.—F, Flute; G, Clarinet; H, Oboe; I, Saxophone

PLATE VI

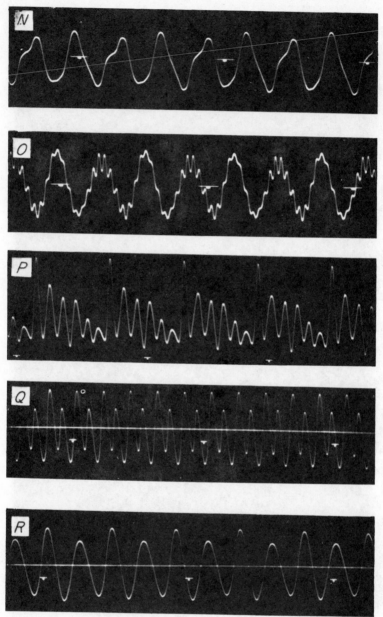

From D. C. Miller's *Sound Waves, Their Shape and Speed* (Macmillan, N.Y.)

FIG. 5.7.—N, the vowel g*loo*m, C.D.H.; O, the vowel b*ee*, C.D.H.; P, the vowel f*a*ther, D.C.M.; Q, the vowel f*a*ther, E.E.M.; R, the vowel n*o*, D.C.M.

the South Kensington Museum. The graph of the vibration is first drawn to the proper scale. The graph is then fixed to the base of the instrument, and a tracing point is moved along the graph through one complete vibration. The instrument reads off the amplitude and relative phase of each harmonic component.

To appreciate the true significance of this type of analysis for musical acoustics we must couple Fourier's Theorem with a law enunciated by G. S. Ohm (1787–1854). He asserted that vibrations of the air which are strictly simple harmonic are unanalysable and are perceived by the ear as simple tones, but that all other forms of periodic air vibration can be analysed by the ear, and each harmonic constituent separately perceived, if of sufficient intensity. Thus the ear acts as a practical Fourier analyser,

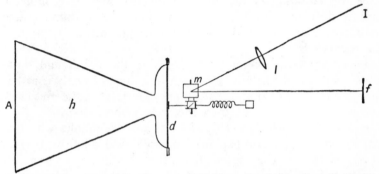

FIG. 5.5.—Miller's Phonodeik. A, open end of horn h. d, thin glass diaphragm. m, small mirror on cylindrical steel staff rotating in jewelled bearings. I, source of light from which beam is focused by lens l, on to mirror m, and reflected to moving film, f

and harmonic analysis in the case of musical notes derives its importance from this fact. What we did when we identified the existence of partial tones by listening to them was to make, by ear, a harmonic analysis of the note.

The Phonodeik.—To make the analysis from a graph, we must first get the displacement curve which represents the motion of the air. One of the methods of doing this uses the phonodeik. This instrument was designed by D. C. Miller,[1] and its action will be understood by reference to Fig. 5.5. h is a horn which collects the sound-waves and directs them on to a very thin glass diaphragm d. To the centre of this diaphragm is attached a very fine wire, which passes round a spindle capable of rotation in

[1] *Science of Musical Sounds*, Macmillan, 1916.

D

jewelled bearings and is held in tension by a light spring. To the spindle is attached a mirror, m, on to which light is focused from a source, I, and from which the beam is reflected on to the moving film f. Obviously when the diaphragm d is set in vibration by the sound-waves, the point of attachment of the wire will oscillate so as to set the spindle rotating to and fro on its axis in time with the movement of the diaphragm. This rotation will move the mirror so that the reflected spot of light vibrates horizontally on the film. The vertical movement of the film draws this out into the required graph. In this the amplitude of the diaphragm is greatly magnified, a movement of $\frac{1}{10}$ mm. of the diaphragm giving a movement of 400 cm. of the spot of light. It is necessary, of course, to assure ourselves that this graph truly represents the vibration of the air, and the necessary tests indicate that it does so with fair approximation. The graph is now re-drawn to the required scale, put through the analyser, and the results recorded.

There are other methods of obtaining the graph and quite other methods of analysis, but it is the results of analysis rather than the method of analysis which are of interest to the musician, and we shall not consider these methods further.

Partial Tones and Musical Quality.—It was Helmholtz who first attempted a systematic explanation of musical quality in terms of partial tones. He insisted that differences in quality were all capable of explanation in terms of the particular selection of partial tones associated with any note and their relative intensities. He knew the laws of vibration of strings, air columns, &c., and he could make some attempt at analysis by ear and by the use of resonators, but he had none of the modern methods available. Nevertheless he was able to establish some correlation between the adjectives used by musical critics to describe musical quality and the physical make-up of the notes in terms of partial tones. He concludes that simple tones, like those of a tuning-fork mounted on a resonator and wide stopped organ-pipes, have a very soft, pleasant sound, free from all roughness, but wanting in power, and dull at low pitches. Musical notes which are accompanied by a moderately loud series of the lower partial tones up to about the sixth are more harmonious and musical. Compared with simple tones they are rich and splendid, while they are at the same time sweet and soft if the higher partials are absent. In this class he includes the piano, open organ-pipes, and the softer tones of the human voice and of the French horn. In a third class he places narrow stopped organ-pipes and the clarinet, these being supposed to give only the odd-numbered partials and to produce a quality of tone hollow and even nasal. When the fundamental tone predominates, the quality of tone is rich, but when the fundamental is weak, the quality is poor. Finally, when partials above the sixth are prominent, the quality is cutting and even rough. In this class come the notes of bowed instruments, of most reed pipes, and of the oboe, the bassoon, the harmonium, and the human voice.

Recent work on the analysis of musical notes may be said to have begun with the investigations made by D. C. Miller using his phonodeik. Examples of his records are given in Fig. 5.6, Plate V. In the first of the photographs, F, we notice a general resemblance to Fig. 5.4. It was made with a flute in G sounding the note $c = 256$ Hz. We are prepared for the statement that it consists mainly of the note and its octave. It contains also the double octave. Odd-numbered partials are absent, and this, together with the fact that only three even-numbered partials are present in any appreciable intensity, accounts for the mellowness and simplicity of certain tones of the flute. Curve G is a record from a clarinet sounding the same note, $c = 256$. The small kinks, from their crowded distribution, must obviously be produced by high partials, and the expert eye may detect that their distribution means that they are once in phase (where the kinks are most marked), and once out of phase (where they almost disappear), in each vibration. They are therefore consecutive partials. There are about eight of these small kinks to each wave-length of the fundamental, so that presumably they are either the seventh and eighth or the eighth and ninth. Analysis by machine shows that the second of these suppositions is correct, and that the reedy tone-quality is probably due to these relatively strong high partials. The curve H is the oboe. The fundamental tone is very weak, and the fourth and fifth partials contain the greater part of the energy, but there are traces of a series of higher partials, indicated by the long and narrow peaks. The curve marked I is recorded with the saxophone. Analysis shows that it contains a strong fundamental with the complete series of upper partials to the fifteenth at least. The fourth partial is the strongest, and the fifth and sixth have moderate intensity.

One of the remarkable results which emerge from the analysis of these and similar records is the comparative weakness of the fundamental. Take, for instance, a typical case—the results of an analysis of the note $g = 196$—to bring out this important point. It is taken from Seashore,[1] and relates to the note as produced on the violin, on the open fourth string.

Order of partial.	Frequency.	Percentage of energy.	Sensation level above the fundamental.
1	196	0·1	0
2	392	26·0	24·2
3	588	45·2	26·6
4	784	8·8	19·5
5	980	8·5	19·3
6	1,176	4·5	16·5
7	1,372	0·1	1·3
8	1,568	4·8	16·8
9	1,764	0·1	0·6
10	1,960	0·0	x
11	2,156	0·1	1·3
12	2,352	0·0	x
13	2,548	0·2	2·4
14	2,744	0·0	x
15	2,940	0·1	x
16	3,136	0·0	x
17	3,332	1·1	10·4
18	3,528	0·1	x
19	3,724	0·2	2·6
20	3,920	0·0	x

The third column shows the distribution of energy, and we find that the fundamental—the partial tone which gives the pitch and is the only tone we consciously hear—carries only 0·1 per cent. of the energy. The quality is presumably due, then, to the second, third, fourth, fifth, sixth, eighth, and seventeenth partials. The sensation level above the fundamental is shown in the last column, and is a rough indication of the relative loudness, which is better represented by sensation level than by energy. It is only an approximate guide, however, as, owing to the fact that the sensitiveness of the ear varies with pitch, two tones of different frequencies may have the same sensation level and still differ in loudness. By making certain assumptions it would be possible to calculate relative loudness more accurately, but the result would not be worth the effort expended.

The analysis of musical notes may also be made automatically by electronic 'sound spectrographs' which can produce a

[1] *Psychology of Music.*

spectrum either on a paper chart or displayed on a cathode-ray tube. Some of these machines store the signal on a tape before analysis and it is possible to use these to study the change of quality during the transient when the sound starts. It has been shown [1] that the initial transient is the most important identifying element of a musical sound and if it is removed it becomes difficult to recognize an instrument.

According to Jeans,[2] the second partial adds clearness and brilliance, but nothing else; the third partial again adds brilliance, but also contributes a certain hollow, throaty, or nasal quality; the fourth adds yet more brilliance, and even shrillness; the fifth adds a rich, somewhat horn-like quality to the tone, while the sixth adds a delicate shrillness of nasal quality.

These six partials are, of course, all parts of the common chord of the fundamental, but this is not true of the seventh, ninth, eleventh, and higher odd-numbered partials. These add dissonance and introduce a real roughness or harshness.

Nature of Vowel-Sounds.—An important application of the technique of analysis is that in connexion with the explanation of vowel-sounds. Early investigators, for whom analysis was impossible, were divided between the fixed-pitch theory and the relative-pitch theory. When a given vowel sound is sung, does the singer always emphasize the partial of a certain order— i.e., the third, fourth, &c.—whatever its pitch, or does he always emphasize a fixed pitch whatever the order of the corresponding partial may be? For instance, a soprano singing the vowel sound *a* in ' father ' emits a note of which the following is an analysis obtained by D. C. Miller using his phonodeik:

Partial.	Frequency.	Energy, %.
1	308	9
2	616	6
3	924	69
4	1,232	8
5	1,540	5
6	1,848	1
7	2,156	—
8	2,464	—
9	2,772	—

If, now, a bass voice sings the same vowel-sound, of course at a different pitch, will the prominent partial be the partial

[1] Eagleson and Eagleson, *J. Acoust. Soc. Amer.*, Vol. 19, p. 338 (1947).
[2] *Science and Music*, Camb. Univ. Press, p. 86.

of the same order as that of the soprano, or the partial nearest to the same pitch? Is it the relative pitch of the strong partial or its absolute pitch that defines the vowel-sound *ah*? The analysis of the vowel-sound sung by the bass voice at once solves this problem for us. It gives:

Order of partial.	Frequency.	Energy, %.
1	154	1
2	308	3
3	462	1
4	616	1
5	772	12
6	924	66
7	1,078	7
8	1,232	7
9	1,386	1

We see at once that the prominent partial is fixed in pitch. This view is verified by extending the observations to other singers and other notes. Always there is a strong partial in the neighbourhood of the frequency. This result is illustrated by Fig. 5.7, Plate VI. In these photographs we notice that the records P and Q are for the same vowel and yet look entirely different. P has a fundamental frequency 172, and the sixth partial with frequency 1,032 contains 80 per cent. of the energy. Q and R look alike, but are for different vowels. Q has a fundamental frequency 487, and the second partial of frequency 974 contains 96 per cent. of the energy. On the other hand, R is the record of the vowel-sound in n*o*, sung at a frequency 226. Its second partial—frequency 452—carries 69 per cent. of the energy. Like Q, its strong partial is the second, but it corresponds to an entirely different vowel.

Extending his observations to other vowel-sounds, Miller found that each vowel of the series m*a*, m*aw*, m*ow*, m*oo*, had a characteristic frequency, the values being m*a*, 910; m*aw*, 732; m*ow*, 461; m*oo*, 326. For the short vowels of the series m*at*, m*et*, m*ate*, m*eet*, Miller obtained two prominent regions of pitch for each— m*at*, 800, 1,840; m*et*, 691, 1,953; m*ate*, 488, 2,461; m*eet*, 308, 3,100. These results have been generally confirmed by other observers using different methods of analysis, except that *all* the vowels seem to be characterized by two prominent regions of pitch, the higher one being less important in the series of vowels to which Miller assigned only one. A region of pitch in which all partials are strengthened is called a formant, and we may say that for every vowel there are two formants of fixed pitch. This result was anticipated in 1879 by Graham Bell, the inventor of the first practical telephone.

These results of analysis can be checked in a variety of ways. Miller recorded whispered vowels, in which only the formant is audible. Paget [1] analysed his own vowel-sounds by ear, and then arranged an artificial larynx in which the part of the vocal cords is played by a reed and the air is blown past this reed

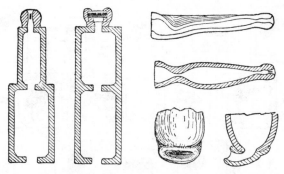

FIG. 5.8.—Sir Richard Paget's models for the production of vowel-sounds

into a double resonator (Fig. 5.8). The two chambers of the resonator are tuned to give the required formants, and in this way the vowel-sound is synthesized. The results for a number of observers have been summarized in the following table due to Fletcher:

Characteristic Frequencies of the Vowel-Sounds

Speech sound.				Low frequency.	High frequency.
u (pool)	.	.	.	400	800
u (put)	.	.	.	475	1,000
o (tone)	.	.	.	500	850
a (talk)	.	.	.	600	950
o (ton)	.	.	.	700	1,150
a (father)	.	.	.	825	1,200
a (tap)	.	.	.	750	1,800
e (ten)	.	.	.	550	1,900
er (pert)	.	.	.	500	1,500
a (tape)	.	.	.	550	2,100
i (tip)	.	.	.	450	2,200
e (team)	.	.	.	375	2,400

[1] *Human Speech*, Kegan Paul, 1930.

It must be remembered that the frequency given is in every case the centre of a range of pitch within which all partial tones are strengthened, and that this range is sometimes considerable. It will be noticed that the vowel sounds *oo* and *ee* involve formants of very low pitch, and this accounts for the great difficulty in producing these vowel-sounds with good enunciation on notes of high pitch. Notes of pitch above that of the formant can have no partial tones lying in the range of the formant, and therefore the appropriate frequency cannot be evoked.

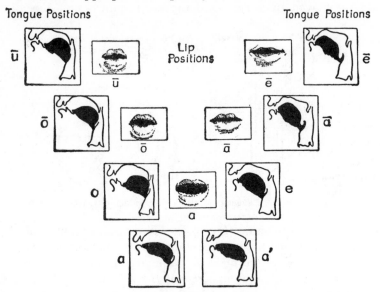

FIG. 5.9.—Tongue and lip positions for two series of vowel-sounds

Recently evidence has been obtained that there are at least five fairly important air cavities associated with voice production. The note produced with its series of partial tones forces these cavities into vibration, and if the natural frequency of any one of these cavities lies near the natural frequency of one of the partials, it is thrown into resonant vibration, and the partial is strengthened. Presumably two of these cavities are important from the point of view of vowel enunciation, and the remainder are effective in other modifications of quality which can be made without affecting the vowel. The two important cavities are probably the mouth and pharynx, their natural frequencies being altered by changing their volume and the width of the

aperture by movements of the tongue and lips. Fig. 5.9 shows
how this is done. The remaining cavities (sinuses, &c.) modify
quality without affecting the enunciation of the vowel. Thus
the vocal cords determine the pitch of the note, two air cavities
determine the vowel, and the remaining cavities determine the
musical quality.

CHAPTER 6

THE EAR

Structure of the Ear.—For convenience of description it is usual to divide the ear into three parts—the outer, middle, and inner ear. The outer ear includes the ' pinna ', the visible, external part. In animals this is often movable, and is useful in discriminating the direction from which a sound comes. It may also act to some extent as a collector of sound-energy. In man, unless the ears are very protuberant, it no longer performs either of these functions, and is at best ornamental, and not always that. Leading from the pinna there is a passage called the ' auditory meatus ', which is closed at its inner end by a very fine membrane, the ' tympanum ' or drumskin. It is the periodic variation of pressure in the meatus, due to the reception of sound-waves which causes the to-and-fro vibration of the drumskin, its vibrations being, of course, executed with the frequency of the pressure changes, and therefore with the frequency of the waves.

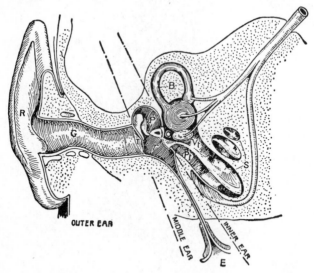

FIG. 6.1.—Diagrammatic section through right ear. R, pinna; G, auditory meatus; T, tympanum; P, chain of bones; O, oval window; S, cochlea; V*t*, *scala vestibuli*; P*t*, *scala tympani*; B, semicircular canals; E, eustachian tube; *r*, round window

78

The middle ear is separated from the outer ear by the drum-skin, to the inner side of which is attached the first of a chain of three little bones or ' ossicles '. These, from their shape, are known as ' malleus ', ' incus ', and ' stapes ', or hammer, anvil, and stirrup. The stirrup is attached, as shown in Fig. 6.1, to the oval membrane and this separates the middle from the inner ear. The middle ear is a cavity completely enclosed except for a connexion with the back of the throat through the ' eustachian tube '. This tube, which is normally closed, but opens during swallowing, serves a double purpose. It acts as a

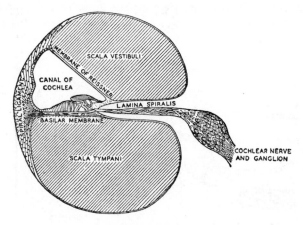

FIG. 6.2.—Cochlea in transverse section

drainage tube and as a pressure equalizer. If the pressure of the air on the outer ear suddenly changes and the pressure in the middle ear remains the same, the difference of pressure on the two sides of the drumskin prevents it from vibrating freely. This happens, for instance, in a rapid aeroplane descent, and sometimes on a sudden entry into a tunnel in a train. In these circumstances we suffer from a temporary deafness, which is relieved at once by the act of swallowing. The function of the ossicles is to transmit the vibrations from the air in the outer ear to the fluid in the inner ear, with suitable modifications of the pressure and amplitude.

The inner ear consists mainly of the ' semi-circular canals ' and the ' cochlea '. The former are associated with our sense of the vertical. The latter is the principal part of the organ of hearing. As its name implies, it is shaped like a snail shell. If

it is opened out and cut transversely, it reveals a cross-section
shown in Fig. 6.2. It is filled with liquid and is surrounded by
rigid bony walls. The oval window is a flexible membrane
closing the *scala vestibuli* at its base, while the round window is
a similar membrane closing the *scala tympani* at its base. As
liquids are nearly incompressible, this arrangement facilitates
the movement of the oval window, as when it is driven in by
the stapes the pressure is relieved by the round window bulging
out, and as the stapes pulls out the oval window, the round
window bulges inwards. In this way vibrations in the fluid are
set up at the oval window, pass up the *scala vestibuli*, cross the
membranes to the *scala tympani*, and return to the round window.

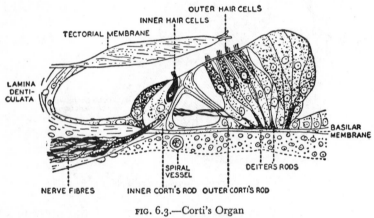

FIG. 6.3.—Corti's Organ

The *scala vestibuli* and the *scala tympani* connect by a tiny passage,
the 'helicotrema', at the apex of the cochlea. The seat of the
stimulation of the nerves is the basilar membrane. This, as is
seen in Fig. 6.3, carries the 'organ of Corti', and this, in turn,
carries the hair-cells. Over these cells lies the 'tectorial membrane'. It is the relative movement of the hair-cells and this
membrane, due to the disturbance of the fluid, which stimulates
the nerves and is responsible for the phenomenon of hearing.
The basilar membrane is about 36 mm long, and tapers in the
reverse direction to the cochlea, being narrowest near the oval
window and widest at the apex.

Sensitiveness of the Ear to Intensity and Loudness.—We have already
seen (p. 35) that there is a lower limit of intensity below which

a sound is inaudible, and its loudness therefore zero. This threshold value of the intensity is shown in Fig. 3.1. At a frequency of 1,000 Hz the threshold intensity for a normal observer is taken as 10^{-12} watts per sq. m. Small as this intensity is, it is not the smallest intensity which can excite the sensation of sound. The ear is most sensitive to sounds of frequency about 3,500 Hz, and for this frequency the threshold intensity is $1 \cdot 55 \times 10^{-13}$ watt per sq. m., the pressure amplitude about $1 \cdot 1 \times 10^{-5}$ Pa, and the displacement amplitude about $1 \cdot 25$ pm.

Let us try to form some conception of the magnitude of these figures. The intensity represents the intensity of the light and heat received from a 50-watt electric lamp at a distance of 3,000 miles in empty space (i.e., if none were being absorbed by the atmosphere). For the pressure amplitude let us remember that the area of the drumskin of the ear is about 1 sq. cm. and the weight of a mosquito is of the order of 1 mg. If, therefore, an insect of weight about one ten-thousandth of the weight of a mosquito were to dance up and down on the drumskin 3,500 times a second, we should hear a sound! Again, the normal pressure of the atmosphere is about a hundred thousand Pa. If this pressure in the outer ear is changed by one part in ten thousand million, and the change is repeated 3,500 times a second, the note would be heard. The necessary change in pressure would be achieved by moving the head up and down through a distance of one micrometre. Finally, the displacement amplitude—the maximum excursion of the air from its mean position during vibration—is only $1 \cdot 25$ pm. This distance is almost unbelievably small. It is about one four-millionth part of the diameter of a fine silk fibre, one four hundred-thousandth of a wave-length of light, one eight-thousandth of the thickness of the thinnest gold leaf, one hundredth of the diameter of a molecule of nitrogen.

As will be seen from Fig. 3.1, the intensity of a sound imposes another limitation on audibility. Not only may a sound be too feeble to be heard, but it may be too intense to be heard. The sensation is then one of feeling rather than hearing, and this limit is called the threshold of feeling. The graph shows how it varies with the frequency of the sound. It is the existence of this threshold of feeling that determines the upper limit of the scale of loudness. For a note of frequency 1,000 Hz the intensity corresponding to the threshold of feeling is 10^{13} times that corresponding to the threshold of audibility. The number of phons of loudness level in this range is 130.

Over the middle of the scale of pitch, and for sounds of average loudness, the smallest increase in loudness which is perceptible is that corresponding to an increase in intensity of about 10 per cent. This means about 0·4 phon. For sounds which are either low or high in pitch and either louder or softer than the average it requires a greater change in loudness to be noticeable. Even for average loudness and medium pitch 0·4 phon can only be discriminated by a good ear under very favourable conditions, and 1 phon may be taken as an approximate value for the ordinary limit of discrimination.

Sensitivity to Pitch.—Just as the eye is sensitive to light only over a comparatively narrow range of wave-lengths or frequencies, so the sensitiveness of the ear to sounds is limited in the same way. There is an enormous range of wave-lengths of electro-magnetic waves from the very long radio waves to the γ rays from radio-active substances—a range of some 50 octaves. Of this range only about 1 octave affects the eye, perhaps fortunately. Radio waves are constantly passing through us in every direction. They are identical with light-waves in all respects except wave-length (or frequency). But we never notice them, and have to use a radio set to detect them.

If we take a source of sound of variable frequency but constant intensity, and gradually increase the frequency from 10 or 15 Hz, we shall at first hear nothing. Then, at a frequency which depends on the intensity we are using, the sound becomes audible as a note of low pitch. Increasing the frequency still further, the pitch continues to rise, until the sound becomes more and more shrill, and then suddenly it vanishes. The vibrations continue, and except in the matter of wave-length they are physically identical with the audible vibrations which have just previously been produced. A glance at Fig. 3.1 will show that the frequencies at which a note becomes audible and ceases to be audible depend on the intensity used, each horizontal line representing a constant intensity. The limits seem to lie at best at about 20 Hz and 25,000 Hz, or about 10 octaves.

The sensitiveness of the ear to notes of high pitch varies in a very marked way with age. Only the young can hear notes of frequency 25,000, and even in middle life the limit is much lower.

A matter of very great interest to the musician is the sensitiveness of the ear to small changes in pitch. This has been measured for pure tones in the laboratory by presenting successively to the ear two tones, one of which can have its frequency gradually increased. When the difference in pitch is just perceptible, the ratio of the frequency of the variable tone to the frequency of the fixed tone measures the just-perceptible interval of pitch. It is found that over the frequency range most common in music (500 to 4,000), and at the usual loudness level (about 40–60 phons), the ear can just discriminate an interval of less than one-thirtieth of a semitone—i.e., about 3 cents. Pitch discrimination is not so good in real life as it is in the laboratory; it is not so good at lower loudness levels; and it is nothing like so good below a frequency of 500, dropping to about one-sixth of a tone, or about 30 cents, at a frequency of 62, even under laboratory conditions. On the other hand, differences of pitch seem much less easy to detect for pure tones than for ordinary musical notes, and Seashore records the fact that among sixteen professionals of the Royal Opera at Vienna great differences were found, the limit of discrimination varying from 1/25th to less than 1/250th of a semitone. The conditions of experiment, however, are not recorded.

Aural Harmonics.—We have already seen (p. 31) that if a mechanism is asymmetric, then, when acted on by a force of frequency f_1, it adds the frequencies $2f_1$, $3f_1$, $4f_1$, &c.—in fact the full harmonic series. There is good reason to believe that the mechanism of the ear behaves in just this way, and that when a pure tone of sufficient intensity falls on it the resulting sensation is that of a note of good quality with its harmonic partials. The existence of these aural harmonics can be shown by introducing simultaneously into the ear (1) a strong pure tone of frequency f, (2) a pure tone of variable frequency and adjustable intensity. If the aural harmonics are generated by the pure tone, then the tone of variable frequency will give beats not only when its frequency is f, but also when its frequency is $2f$, $3f$, $4f$, &c. Also, by adjusting the intensity of the comparison tone while it is giving beats, we can find the intensity for which the beats are most distinct, and this will be the intensity of the aural harmonic. By methods of this kind Fletcher [1]

[1] *Speech and Hearing*, p. 178. Macmillan, 1929.

obtained the curves shown in Fig. 6.4.   The curves show, for any frequency, the intensity level for the pure tone at which each harmonic appears.   The phenomenon is set out in a rather different way in Fig. 6.5.   For any given intensity of the fundamental (first harmonic) the intersection of the corresponding line with the ordinates gives the intensity of the other harmonics. Thus, if the fundamental has an intensity level of 100 dB, the other harmonics have intensity levels 88, 74, 61, 48, 37, 28, 21, 14

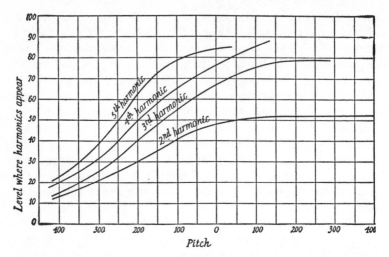

FIG. 6.4.—Sensation levels at which aural harmonics become perceptible.   The pitch is measured in hundredths of an octave from a pitch of 1,000 Hz, which is taken as zero.   The sensation level is in decibels above the minimum audible

dB.   If the intensity level of the fundamental is 80 dB, the other harmonics have intensity levels 62, 45, 32, 20, 10, 0, 0, 0 dB.   In actual practice 60 dB would be a more usual intensity level for the fundamental, and in this case only the second and third harmonics would appear at levels of 30 and 12 dB respectively.

Since the contour lines of equal loudness for intensity levels of 60 dB and over are nearly horizontal from a frequency of about 300 to about 5,000 the above figures for intensity level are a fair indication of loudness.

Aural Combination Tones.—We have seen (p. 30) that the effect of the double forcing of an asymmetric system is to produce combination tones.   Now, the ear is such a system.   We should therefore expect that combination tones would be formed in the ear, and would, together with the aural harmonics, profoundly modify the resulting sensation. Sometimes it is found that the combination tones are strengthened when an appropriate resonator is placed to the ear. In this case they must be produced externally, either in the instrument or in the air. Frequently,

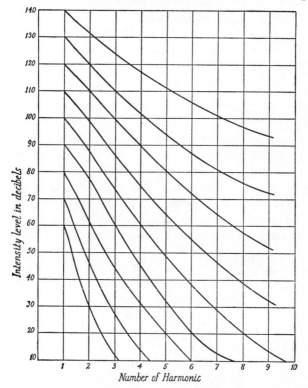

Reprinted by permission from *Hearing : Its Psychology and Physiology*, by Stevens and Davis, published by John Wiley & Sons, Inc.

FIG. 6.5.—Relative intensities of aural harmonics. The starting-point of a line on the abscissa 1 gives the intensity level of the fundamental, while the intersection of the line with the other abscissae gives the intensity levels of the corresponding partials

however, the combination tone can be heard without a resonator and is not strengthened when the resonator is used. In this case the combination tone must be produced inside the ear itself.

The existence of these tones has been established by Wegel and Lane (Fletcher, *Speech and Hearing*, p. 176), using the method of the exploring tone as in the case of aural harmonics and noting the beats. Using pure tones of frequencies 700 and 1,200, they detected the following frequencies:

(a) Harmonics of 1st tone—700, 1,400, 2,100, 2,800.
(b) Harmonics of 2nd tone—1,200, 2,400, 3,600.
(c) First-Order Difference Tone—500.
(d) First-Order Summation Tone—1,900.
(e) Second-Order Difference Tones—200, 1,700.
(f) Second-Order Summation Tones—3,100, 2,600.
(g) Third or Higher Order Tones—4,300, 3,800, 3,300, 2,900, 1,000, 900.

If we put $f_1 = 1,200$ and $f_2 = 700$, then these last frequencies are $3f_1 + f_2$, $2f_1 + 2f_2$, $3f_2 + f_1$, $3f_1 - f_2$, $2f_1 - 2f_2$, and $3f_2 - f_1$.

These results may explain one or two facts of observation which seem at first rather surprising. Analysis shows that a source of sound may sometimes radiate 1 per cent. or less of its energy in its fundamental, and still the impression of the listener may be that he hears one note only—the fundamental. It is to be noticed, however, that if the source is producing the harmonic series of partial tones, the frequencies of these will be f, $2f$, $3f$, $4f$, &c., and the difference of frequency between each successive pair is f. Thus each successive pair of partials combines to give a difference tone of the frequency of the fundamental, and therefore to strengthen the impression which the fundamental makes. This explanation is borne out by experiments by Fletcher,[1] who used ten vacuum-tube generators to give pure tones of the first ten members of the harmonic series 100, 200, 300, &c. When all these tones were present, a full tone was heard of frequency 100. When the tube giving this note was cut out, very little change was observed; the impression was still that of a pitch corresponding to a 100 Hz tone. Even with the first seven harmonics eliminated, leaving only frequencies of 800, 900, and 1,000, the pitch was that of a 100 Hz tone. Any three successive harmonics produced the same effect. Some recent experiments [2] suggest that these results exaggerate the effect of the difference tone in determining pitch, but others fully confirm the conclusions which Fletcher drew.[3]

Two sets of gramophone records have been produced by the Bell Telephone Company. The two sets are identical, except

[1] *Journ. Acous. Soc. Amer.*, Vol. 6, p. 59 (1934).
[2] *Ibid.*, Vol. 13, p. 84 (1941).   [3] *Ibid.*, Vol. 13, p. 87 (1941).

that in one all the frequencies below a certain limit are filtered out. The two sets sound very much the same, speech and music being perfectly intelligible, although all the fundamental tones are missing except in the form of difference tones created by the ear of the listener.

In the earliest form of gramophone there was a small horn whose function was not understood. Actually it was, in effect, a high pass filter, cutting out all frequencies below a certain value. Fortunately the ear supplied the missing bass by the creation of difference tones. Exactly the same is true of some radio sets which cut out all frequencies below about 250, and so transmit no bass or tenor tones at all. A *pure* tone of frequency 250 is unheard, so also is a note with partials which are not harmonic, like the sound of the drum or of any other percussion instrument. For the same reason, applause at the end of a concert item is entirely changed in character.

Theories of Hearing.—It is interesting to consider the facts of musical perception alongside of the known structure of the ear, and see how far it is possible to construct a theory which will explain the facts in terms of the structure. A most attractive theory is that which seems to have been first suggested by Cotugno (1736–1822) and later by Charles Bell (1774–1842). It was developed in great detail by Helmholtz, and is known as the resonance theory. On this theory the fibres of the basilar membrane which are transverse and in tension act like resonators fairly heavily damped.

The important facts to be explained are:

(1) the ear responds to pure tones over a range of frequency of ten or eleven octaves;

(2) pitch discrimination is acute over the middle of the range of audibility, but becomes less acute for notes of high pitch and much less acute for notes of low pitch;

(3) when two tones of nearly equal frequency are sounded simultaneously, we hear beats, but only when the musical interval between the two tones is not too large;

(4) trills are clear-cut and distinct if not too rapid, but become confused if the rapidity is too great, the admissible rapidity being greater in the treble than in the bass;

(5) two or more notes sounded simultaneously can be separately perceived, as can the constituent partial tones of a musical note.

If we now consider the possible explanations of these facts on the Helmholtz theory we may note the following points.

(1) Competent anatomists have maintained that, taking into account the variation in length and tension of the fibres of the basilar membrane and their loading by the liquid content of the cochlea, they can cover the required range of pitch. Since the fibres are damped, resonance will not be sharp. A group of fibres will be excited by a pure tone, and the position of the fibre whose stimulation is greatest will determine the pitch of the tone.

(2) Over the middle of the range it is reasonable to suppose that two maxima can be distinguished as separate, even when fairly close together. The number of fibres seems to make the actual observed capacity for discrimination possible. On the other hand, near the ends of the range the resonators will be stimulated by tones whose frequencies lie beyond the proper frequencies of the end resonators, and no true maxima exist.

(3) Beats occur when two different tones act on the same resonator. The effect is well shown if two forks of slightly different frequency are sounded and their shafts put on the base of a monochord on which is stretched a wire giving a note of intermediate frequency. The vibrations of the wire will increase and decrease periodically. The resonance theory suggests that the transverse fibres of the basilar membrane can act like the wire. If two notes are sufficiently close in pitch for the two groups of excited fibres to overlap, then the fibres constituting the overlap will be the seat of the beating. If the tones are too widely separated in pitch for the corresponding groups of fibres to have any members in common, beats cannot be heard.

(4) The fact that clear and distinct trills can be produced at all bears out the suggestion that the fibres are damped, and the degree of damping required fits reasonably with other requirements. Also the damping is measured by the amount which the amplitude of the vibration falls off in a given number of vibrations, and not the falling off in a given time. As vibrations take longer in the bass, confusion ought to be greatest there.

(5) The various sensitive nerve-endings distributed along the basilar membrane have separate nerve-connexions, and therefore if a series of notes stimulate a series of groups of fibres, each group can produce its own mental effect.

Whether the fibres do act in this way is still a matter of dispute. The theory is given in detail because at least it serves one important purpose of a true theory—it gives a picturesque representation of an important series of facts, making them coherent and easily intelligible. One thing is hardly in doubt: sensitiveness to pitch is distributed along the basilar membrane, sensitiveness to notes of high pitch being located where the fibres are short—i.e., near the oval window—and sensitiveness to notes of low pitch where the fibres are long—i.e., at the end remote from the oval window. In the case of animals having the same kind of organ as man—e.g., the cat and the guinea-pig—direct experiment shows the distribution of pitch sensitiveness. If animals are subjected to continuous intense sound of definite pitch and the cochlea subsequently examined, localized damage is found in the basilar membrane, and the position of this damage depends on the frequency of the note. Also localized damage can be produced in the cochlea of a guinea-pig by inserting a very fine drill, and the animal is afterwards found to suffer from a selective deafness to tones of a particular pitch only.

There are at least two rivals to the resonance theory and the interested reader is referred to more specialized texts, for example *Theory of Hearing* by E. G. Wever, for further discussion.

CHAPTER 7

VIBRATIONS OF STRINGS

PROBABLY the most important class of musical instruments is that based on the vibrating string, plucked, struck, or bowed. There is reason to believe that it has its origin in the hunting-bow. When the arrow was released the musical note of the bow-string must have been fairly obvious, and there is evidence that the bow was used as a musical instrument as well as a weapon. The musical use of stretched strings was familiar in Old Testament times. The harp is frequently mentioned, as are special kinds of harp, the lute, the psaltery, the viol, and the dulcimer. The Greek tetrachord played an important part in the development of the musical scale, and Pythagoras (572–497), as we have seen (p. 50), used the vibrating string to study the relation between the notes forming the simpler musical intervals.

In the science of acoustics, as in almost every other branch of science, very few discoveries of any importance were made during the 2,000 years that followed the time of Pythagoras, and we find the next advance being made by Galileo (1564–1642). He discovered the quantitative relation between the frequency of the vibration and the length, diameter, density, and tension of the string. This investigation was made independently by Mersenne (1588–1648) of Paris, and the results were published by him in 1636. He found that the frequency of the stretched string is

    (1) inversely proportional to the length,
    (2) directly proportional to the square root of the tension,
    (3) inversely proportional to the diameter for wires of the same material,
    (3a) inversely proportional to the mass (and therefore the weight as ordinarily measured) of unit length of the string.

These laws are qualitatively illustrated in the everyday use of stringed instruments. On the violin, the scale is produced by stopping with the fingers so as to vary the vibrating length. In the piano, as in the violin, to tune the strings the tension is altered by screwing the peg to which the string is attached. To secure a low note with the string in sufficient tension to transmit the vibrations to the radiating surface of the instrument without using an unduly long string or an unduly thick one, the mass may be increased by wrapping fine metal wire round a gut string (as in the case of the violin G string).

Mersenne's Laws may be illustrated quantitatively by the use of a sonometer. Two or more wires are mounted on a strong frame. They are fixed to pegs at one end, pass over a fixed bridge, then over a movable bridge, and finally over a pulley. A scale-pan of known weight is attached to the end of either wire.

(1) A suitable tension is applied by placing weights in the scale-pan. The movable bridge is then adjusted so as to tune the vibrating portion of the wire to a fork of frequency about 256. This can be done by ear, or alternatively by using the principle of resonance. In the latter case a paper rider is placed on the wire, the sounding-fork is held with its shaft on one of the bridges, and the adjustable bridge is moved till the response of the rider is a maximum. In either case the adjustment is made for a series of forks—perhaps the octave series starting from the 256 fork—and the tuned length noted in each case. If the first law quoted above holds, then the product of the frequency of a fork and the vibrating length will be the same for all the forks. Or we can plot a graph for our observations, plotting frequency against the reciprocal of the length, and we shall obtain a straight line.

(2) Select two forks whose frequencies are in the ratio of say $5:4$— i.e., two forks forming the interval of the major third. Adjust the vibrating length of wire to unison with the lower fork for a moderate tension. Note the weight W_1 in the pan. Let the weight of the pan itself be w. Now add weights without altering the vibrating length until the wire is in unison with the higher fork. Let W_2 be the weight in the pan. It will be found that

$$\frac{W_2 + w}{W_1 + w} = \frac{5^2}{4^2} = \frac{25}{16}.$$

(3) Fix to the apparatus two wires of different thickness and material. Attach the scale-pan to the first and add a suitable weight. Adjust the length of wire till it is in unison with a given fork. Let the length be l_1. Now transfer the scale-pan with weights to the other wire and adjust the length until the wire is in unison with the same fork. Let this length be l_2. Let the frequency of the fork be f. Then if the second wire were identical in material and thickness with the first, its frequency would, from Mersenne's first Law, be

$$f \times \frac{l_1}{l_2}$$

Weigh a known length of each wire and calculate the weight of 1 m.

of each. Let the weights be m_1 and m_2 respectively. Then, if Mersenne's third Law holds, the frequency of the second wire will be

$$f \times \frac{l_1}{l_2} \times \sqrt{\frac{m_1}{m_2}}$$

But its frequency is f

$$\therefore \frac{fl_1}{l_2}\sqrt{\frac{m_1}{m_2}} = f$$
$$\therefore \ l_1\sqrt{m_1} = l_2\sqrt{m_2}$$

Quality of Notes from Stretched Strings.—There are three ways in which the quality of the note given by a stretched string may be varied:

(1) by variation of the point of attack,
(2) by variation of the method of attack,
(3) by variation of the vibrating system to which the wire is coupled.

The ideal string—a perfectly flexible string between rigid supports—gives the harmonic series of partials. The actual string shows a departure from these ideal conditions, and the combined effect of the stiffness of the string and the yield of the supports is to give a series of partials all sharper than the corresponding member of the harmonic series, the departure being greater for short strings than for long ones and greater for high harmonics than low ones.[1] These inharmonic partials are presumably more noticeable in the free vibrations of the string, the notes of the pianoforte, or the pizzicato notes of the strings, than in the bowed notes of the strings. The importance of the point of attack is that it can be selected to diminish, if not to cut out entirely, whole groups of partials.

There are three distinct ways in which a stretched string can be set in transverse vibration, and each way is associated with a specific quality of tone. The string may be plucked, as in the harp; struck, as in the pianoforte; or bowed, as in the violin.

As we shall see in discussing particular instruments, the quality of tone is very greatly dependent on the associated vibrators. The sound-board of the piano is almost the most important thing about it, while the differences in tone given by various violins depend on the vibrations not only of the body of the instrument, but of the contained air as well.

The Pianoforte.—As an illustration of sound-production by the striking of a stretched string or wire, we may consider the piano-

[1] Shankland and Coltman, *Journ. Acous. Soc. Amer.*, Vol. 10, p. 161 (1939).

forte. Its characteristic features are its series of strings, its hammer action, and its keyboard. Its predecessors were the clavichord, the spinet, and the harpsichord. The clavichord key was a simple lever, one end of which was pressed by the finger, while the other end extended under the strings of the instrument and carried an upright piece of metal. When the key was depressed, the piece of metal, called the tangent, was forced up so that its edge caught the string. In relation to the string the tangent fulfilled the double purpose of (*a*) throwing it into vibration by acting as a hammer, (*b*) determining the vibrating length of the string by remaining in contact and acting as a bridge. In this way it divided the string into two segments, in one of which the vibrations were damped by strips of cloth intertwined with the strings. With this instrument some

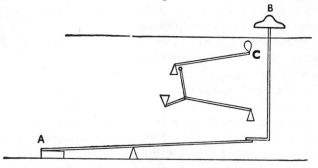

FIG. 7.1.—Simplified pianoforte action with key A depressed, damper B lifted, and hammer C about to strike the wire

variety of quantity and quality of tone was possible by adjusting the striking force, but at best the loudness of tone was quite inadequate. On the other hand, in the spinet and the harpsichord the strings were plucked, and although greater loudness could be obtained, no variation in loudness or quality was possible. It was in these circumstances that in 1709 Bartolommeo Cristofori exhibited specimens of harpsichords with hammer-action capable of producing *piano* and *forte* effects (hence the name of the new instrument). The instruments were shown to Prince Ferdinando dei Medici, and in 1711 a full description was published. The principal changes which have taken place since the time of Cristofori are in the action and the range ($7\frac{1}{3}$ octaves as against $4\frac{1}{2}$ octaves), and this latter change has been partly due to the introduction of the steel frame, which is capable of sustaining a very much greater tension than it had previously been possible to use. The action, shown in Fig. 7.1, is devised so that when the key is depressed a hopper is raised which exerts a thrust against the hammer-shank. This throws the hammer-head against the string. On recoil the hammer is caught by a check, holding it ready for repetition. Depressing the key raises a damper from the corresponding string, while releasing the key restores the damper. The left pedal shifts the action so that the hammers strike one of the double or two of the triple strings or, in upright pianos, moves the hammer action nearer the strings, so that the hammer-heads describe shorter arcs. This is the ' soft ' pedal. The right-hand, or ' loud ', pedal removes all the dampers from the strings. This not only gives greater loudness, but also enables the strings not actually being struck to vibrate in unison with the partials of the sounding strings.

The hammer consists of a head and shank of wood, the head being covered with a thin strip of leather and several layers of felt. The hardness of this felt influences the quality very greatly. Hard, compressed felt throws more energy into high partials, while soft, teased felt gives a strong fundamental and a strong second partial, but throws relatively little energy into the higher partials.

The quality of the note given by a stretched string will depend, as we have seen, on the point of attack. A theoretical treatment of the problem shows that the proportion of the energy of the hammer-blow which goes into a particular partial tone depends on two factors, one determined by the method of attack and one by the point of attack. If the point of attack coincides with a node for a particular partial, that partial will tend to be absent or at minimum intensity, while if it coincides with an antinode, the partial will be present with great strength. The actual point of attack chosen for the pianoforte-hammers had, of course, been decided empirically long before any theoretical treatment was possible. It seems to vary round about the fraction 0·12 of the length from one end. The dissonant partials are the seventh, ninth, and eleventh, and their nodes are at 0·14, 0·11, and 0·09. Thus the position chosen is a good one for weakening these partials. It is claimed, however, that the position chosen has other advantages. It is near the point which gives maximum efficiency (maximum transfer of the energy of the hammer to the string), maximum fullness of the fundamental, and a maximum forced vibration of the sound-board.

The strings vary from about 2 m in length in the bass to about

50 mm in the treble.   In the modern grand piano we may have, beginning in the extreme bass,

    (1) 8 single wires wound with a wire of an alloy,
    (2) 5 pairs of wires in unison, covered,
    (3) 7 sets of threes in unison, covered,
    (4) 68 sets of threes in unison uncovered.

This gives 243 strings for 88 notes.   The total tension on the frame may be as high as 30 tons weight.   In the old wooden-frame pianos tensions of this magnitude could not have been used, and with lower tensions the strings were shorter and quality poorer.   The introduction of the steel frame made these high tensions available, and as the steel frame did not readily warp, the need for frequent tuning became less.   The strings are stretched over two bridges, one carried by the frame and one by the sound-board.

The function of the sound-board is to modify string quality and to radiate the sound.   Its vibration is forced by the string, and it is important that as far as possible the whole board should vibrate together.   Now the vibration is applied at the bridge, and if it is to spread quickly to all parts of the board, the velocity of the waves through the board must be high.   It can be shown that this is secured by selecting a wood with high elasticity and low density, such as Norway spruce.   Along the grain in this wood the velocity is about 5,000 m per second. Across the grain it is less, but is increased by the fixing of ribs which run in this direction and increase the stiffness, and therefore the velocity.

Originally the pianoforte had its strings horizontal, as in the modern ' grand '.   The introduction of the upright, or cottage, model made it possible to have one in every home, and had a great deal to do with the development of home music.   But, in spite of many virtues, the pianoforte has some startling defects.   The most obvious is its failure to sustain a tone.   Like all percussion instruments, its tone reaches a maximum just after the moment of impact and immediately begins to die away.   This means a beautifully crisp and sharp attack, but no subsequent control.   There is no reason why a pianoforte string should not be maintained electrically and the loudness of its tone varied at will while it is so maintained. This would give the pianist marvellous control and all the advantages which go with it.   Perhaps, however, we are thinking too much in the terms of the physicist.   Music is probably at its best when it is played on an instrument similar to that for which it was written, just as it is best when performed in the kind of building for which it was written. If we improve the instrument out of all recognition, we shall have to wait for a new generation of composers to take advantage of the opportunities it offers.

Pianoforte Touch.—This brings us to the vexed question of touch. Pianists have generally maintained that the way in which the note is played—i.e., the way in which the key is depressed—affects the quality of the note produced, and that quality and loudness can be separately controlled. Physicists have generally denied this. They have pointed out that before the key has reached its bed the hammer is no longer in contact with the mechanism. No amount of waggling, vibrating, rocking, or caressing affects the hammer or the tone after the instant of release. In fact, the only thing the player controls is the velocity of the key at the instant at which the hammer is thrown off. Thus the pianist can produce indirectly a great variety of tone qualities, but only by control of the intensity of the tone. This view is strongly supported by recent work.[1] A mechanical striker was used to strike the key in a great variety of known and accurately controlled ways imitative of different methods of touch. The vibration of the string was recorded. It was found that (*a*) the vibration depended only on the speed with which the hammer struck the string, and was quite independent of the method of touch, and that (*b*) any vibration produced by a skilled pianist could be exactly reproduced by the mechanical striker. Two pairs of records are shown in Fig. 7.2. On this view the ' richness ' or ' mellowness ', &c., of a tone produced by a novice cannot be improved upon or altered in any way by a skilled performer playing the same note on the same piano, unless he chooses to alter the loudness. The skill of the pianist resides in the accuracy of his control of hammer-velocity, and the method of touch affects this factor very much. His variation in timing and in intensity give the quality to a passage.

The same workers have shown the very great differences in quality which are associated with differences in loudness. An example is seen in Fig. 7.3. No quantitative exactness is claimed for these records, but the relatively greater importance of the higher partials in the case of the fortissimo note is very noticeable. A difference of quality is also found between bass and treble notes when struck with the same force, the fundamental being relatively much weaker in the bass.

[1] Hart, Fuller, and Lusby, *Journ. Acous. Soc. Amer.*, Vol. 6, p. 80 (1934).

Bowed Instruments.—As an illustration of the production of music by bowing a stretched string or wire, we may consider the violin. It is the principal member of the most important group of orchestral instruments. Dealing with the violin as representative also of the viola, the violoncello, and the double bass, we may describe it as a

FIG. 7.2.—Records of notes struck by pianists and by mechanical striker, traced from photographs given in the *Journal of the Acoustical Society of America*, Vol. 6, p. 90 (1934)

resonant box of peculiar shape closed except for two *f*-shaped holes on the top or belly. Over a bridge supported on the belly four strings are stretched, one end of each string being attached to a tail-board and the other to a peg by means of which a tension can be applied. The vibrations of the string are transmitted by

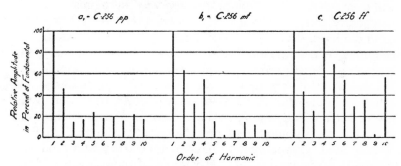

FIG. 7.3.—Harmonic analyses of wave-forms for the sound given by the same note struck to give three different degrees of loudness

the bridge to the belly, and thence to the back, partly through the sides, but largely through the sound-post, a rod nearly under the right foot of the bridge connecting the belly and the back. Under the left foot of the bridge is a bar glued to the underside of the belly, and known as the bass bar. It runs longitudinally, distributing the pressure and transmitting the vibrations more rapidly, as in the case of the bars on the pianoforte sound-board.

The early history of the violin is difficult to unravel. The first bowed instrument seems to have been the *tromba marina*, a curious instrument with a lop-sided bridge which vibrated on the belly, producing a very characteristic tone quality. The single string was played with a bow and different notes were obtained by touching the string with a finger *between* the bow and the bridge. The technique is somewhat difficult to acquire and a considerable amount of luck is needed to ensure that the bridge remains in the proper position throughout a performance. The *tromba marina* is a form of monochord, an oblong box with a triangular nut at each end. Under the string was a scale giving fractions of the length. There was a movable bridge used for adjusting the vibrating length of string. This bridge, movable but no longer moved, survives in the violin. The tension of the string was first adjusted by weights, but afterwards by screw-pegs. Another early ancestor of the violin was the *crwth* or *crowd*, itself a descendant of the Greek lyre. The *rebec* or bowed instrument, with three strings and a pear-shaped body with sound-holes, appeared at latest in the thirteenth century, and probably very much earlier. The immediate ancestors of the violins were the viols, whose features are preserved in the double bass, with its flat back, its high bridge, slanting shoulders, and deep sides. The violin proper appeared in the middle of the sixteenth century, and by 1571 there were seven violins in the royal band of Queen Elizabeth. The home of the perfected violin was Cremona. Andreas Amati introduced some improvements in the instrument, but the contribution made by his sons Antonio and Geronimo, who were partners in a violin-making business, was still more important. Perfection was reached in the workmanship first of Nichola Amati (1596–1684), son of Geronimo, and second of the famous Antonio Stradivari (1644–1737), who fixed the final form of the Cremona model. Guiseppe Guarneri also produced instruments remarkable for the boldness of their design and for the power of their tone.

Process of Bowing.—The advantage of the violin over the piano is its capacity to sustain tone. This is achieved by the process of bowing, which was investigated by Helmholtz. The bow draws the string aside, the tension of the string supplying a gradually increasing force, tending to make the string slip past the hair of the bow. When the friction can no longer hold the string against the bow, slipping takes place, and the string slips back to its undisplaced position and beyond. Presently it comes to rest, is again caught by the bow, and again displaced until its slips once more.

It is possible to study the motion of the string by a recording device. A narrow slit illuminated from behind is placed at right angles to the string and just in front of it. Looking at the slit, we see a dark spot on it which is the string at the point where it crosses the slit. If now the string is set in vibration, the spot moves up and

FIG. 7.4.—Vibration curve for a bowed string, showing a constant speed of displacement with the bow and a different constant speed of slip-back against the bow

down the slit, and if the slit is photographed on a moving film, we can see the exact form of the vibration. The graph obtained is in the form of Fig. 7.4, and consists of pairs of straight lines. This suggests that the point first moves with a uniform velocity (that of the bow) in one direction, and then with a uniform but generally a different velocity slips past the bow in the opposite direction. That it does indeed move with the bow without any slip when it is being displaced can be proved by recording the motion of the bow on the same film. The slopes of the two lines are proportional to the

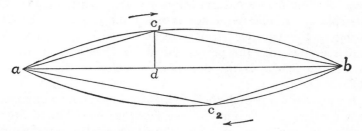

FIG. 7.5.—Successive forms assumed by a bowed string are given by the diagram if d is assumed to move from a to b and back with constant speed while the end of the perpendicular from it moves round the arcs ac_1b, bc_2a

speeds, and depend on the point observed. If the point is at the centre the two lines have the same slope, and the point on the string moves forward with a given speed, stops, reverses, and returns with the same speed. If the point is not at the centre, the two speeds are in the ratio of the two lengths into which the point of observation divides the string.   Perhaps the motion of the string can be more clearly understood from Fig. 7.5, which is given by Helmholtz.   The foot d of the ordinate of the highest point moves backwards and forwards with a constant velocity on the horizontal line ab, while the highest point of the string describes in succession the two arcs ac_1b and bc_2a.   All points on the string pass through their undisplaced positions simultaneously.   The force between the bow and the string is a little greater as the bow draws the string aside than it is as the string slips past the hair.   This is due to the fact that it is always takes more force to start a body moving against friction than to maintain it in motion once slipping has started. Thus the bow does a little more work on the string in displacing it than the string does on the bow in slipping past it, and this balance of work provides the energy for radiation as sound.   It is interesting to notice that all the music produced by bowed-string instruments depends on the very small difference between the maximum force of friction when string and bow move together and the force when the string is slipping past the bow.

The ' Wolf' Note.—The bowed string is an interesting case of forced vibration.   The force—due to the bow—has no natural frequency of its own, so that the system whose vibration is forced— the string—vibrates with its own natural frequency.   The violin is, of course, a complicated coupled system.   The bridge, the belly, the back, and the contained air all take part in the vibration. Each of these has its natural frequency at which resonance may take place.   If this resonance is sharp the effect will be bad, owing to the selective reinforcement of particular notes.   We shall see later that the proper distribution of these natural resonant frequencies is the determining factor in the quality of violin tone.

A sharp resonance of the belly of the violin has another important consequence.   It accounts for the fact that most bowed instruments have one note which is difficult to produce satisfactorily and which tends to give a ' howling ' quality.   It is called the ' wolf' note.   For any given rate of loss of energy from the string, a certain minimum bowing pressure is required to maintain the vibration in its usual form with the fundamental predominant.   Investigations by Raman [1] showed a marked variation of this critical bowing pressure with pitch.   Fig. 7.6 illustrates this point and shows two

[1] *Proc. Ind. Ass. for the Cultivation of Science*, Vol. 6, p. 19 (1920–1921).

PLATE VII

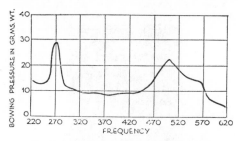

FIG. 7.6.—Relation between minimum bowing pressure required to maintain the vibration of the string in its usual form and the frequency

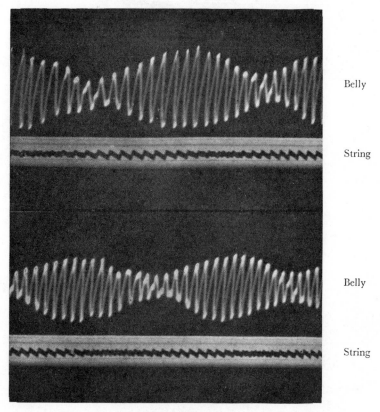

FIG. 7.7.—Time axis. Simultaneous vibration-curves of belly and string of violoncello at the 'wolf-note' pitch

PLATE VIII

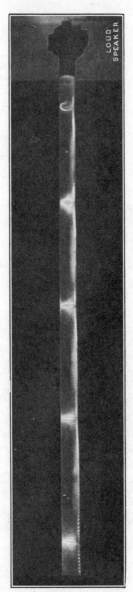

FIG. 8.3.—Kundt's Tube.   The discs which form at the antinodes can be clearly seen

maxima, one of them a very sharp one. The sharp maximum at 270 coincides with the natural frequency of the air contained in the instrument. The maximum at 500 is one of the resonance frequencies of the belly, and was found to be the pitch of the wolf note. As the string is coupled to the belly of the instrument through the bridge and its natural frequency coincides with one of the natural frequencies of the belly, this latter will rapidly increase its amplitude, taking energy from the string until the bow is no longer able to maintain the ordinary form of vibration. The string then changes to a vibration in which the second partial (the octave) is predominant. The belly is no longer in resonance with this new frequency, its amplitude diminishes, the rate at which it takes energy from the string diminishes until presently the bow takes charge again and the cycle repeats itself. That this is what actually happens is clear from Fig. 7.7, which shows simultaneous photographs of the vibrations of the string and of the belly of a violoncello. The swelling and lulling of the vibrations of both string and belly are clearly shown, as is also the fact that when the vibrations of the string are a minimum (just after the vibrations of the belly are a maximum) the peaks on the record of the string vibrations are twice as close, and so reveal the fact that the frequency of vibration is twice as great. This explanation is further borne out by the fact that when the belly is ' loaded ' by adding the mute to the bridge which is already resting on it, the load increases its inertia and diminishes its natural frequency, and the wolf note is found to have dropped in pitch. As has been already pointed out, the sharper the resonances of the belly of the instrument the more pronounced is the wolf note and the greater is the bowing pressure required to control it.

Quality of Violin Tone.—It is found that quality of tone is affected very little by the bowing pressure, but more by the speed of the bow. The factor which is most important, however, is the point of attack on the string. If the bow were extremely narrow the tone of the string would contain very prominent high partials and the quality of tone would be independent of the point at which the bow was applied. Owing to the considerable width of the bow, this is not true, and the nearer to the bridge the point of application of the bow the more brilliant the quality. The extreme positions for bowing ordinarily used are about $\frac{1}{5}$th to about $\frac{1}{25}$th of the length of the vibrating string from the bridge, the more usual limits being from $\frac{1}{7}$th to $\frac{1}{15}$th, and the

E

common practice is to choose a position at about $\frac{1}{9}$th or $\frac{1}{10}$th of the length of the vibrating string from the bridge. This tends to eliminate, or at least to subordinate, the dissonant partials, to which reference has already been made in connexion with the pianoforte. The device of playing very close to the bridge (*sul ponticello*) is a well-known one for achieving brilliant quality. It has to be accompanied by higher bowing pressure, since the critical bowing pressure required to control the vibrations of the string increases as the point of application of the bow gets nearer to the bridge. The changes in quality produced in these ways, however, are none of them very great compared with the varia-

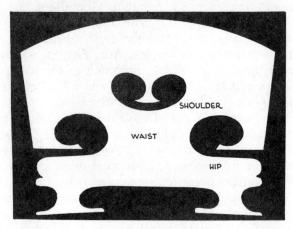

FIG. 7.8.—A violin bridge

tion involved as we pass from one note to another. The differences in quality between the same note played on two different strings or between two different notes played on the same string is one of the remarkable features of the violin.

The intensity of the sound produced varies a little with bowing pressure and with point of attack, but depends mainly on the speed of bowing. The range of intensity is from 25 to 30 decibels.

The bridge transmits vibrations from the strings to the belly. It seems to be designed to eliminate the longitudinal vibrations of the string (those in the line of the string) by the comparatively free vibration of the top part about the waist-line (see Fig. 7.8). The right foot of the bridge, being almost over the sound-post, moves very little, so that the bridge rocks about this foot in its own plane, the left foot setting the belly in vibration, and these vibrations being communicated from point to point in the wood of the instrument and from the wood to the contained air. Reducing slightly the width of the shoulders by filing diminishes the intensity of the high partials and modifies the quality.

The effect of the mute is not merely to give less loudness. It makes profound changes in quality by lowering the natural frequencies of the instrument. Thus for the lower tones it tends to strengthen the fundamental at the expense of the partials. This effect on the natural frequencies of the instrument is clear if we compare Fig. 7.9 with Fig. 7.6. They are obtained on the same instrument and under the same conditions, except that the bridge carries a mute weighing 12·4 gm. The lowest maximum,

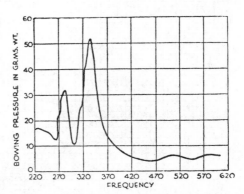

FIG. 7.9.—Relation between minimum bowing pressure and frequency (with mute)

at 270, is practically unaltered, which is what would be expected if it were due to vibrations of the contained air, as these would not involve much motion of the bridge and the additional load provided by the mute. On the other hand, the maximum at 500 in Fig. 7.6 drops to about 330 in Fig. 7.9, and is greatly accentuated. The drop in pitch is due to the fact that we are here concerned with a vibration of the belly, and therefore of the bridge, which is slowed down by the additional load. This lower pitch becomes the pitch of the wolf note.

An exhaustive study of the violin has been undertaken by F. A. Saunders,[1] especially with a view to comparing the performance of the old and famous Cremona instruments with those of modern production. About a dozen of the very best old instruments have been studied and compared with some thirty modern ones. In making a comparison of this kind it must be remembered that few of the old instruments can now be as they were when they left the hands of Stradivari and his fellow-

[1] *Journ. Acous. Soc. Amer.*, Vol. 9, p. 81 (1937); and *J. Frank. Inst.*, Vol. 229, p. 1 (1940).

workers. Many of them have had new and thicker bass-bars to withstand the increased tension of the strings due to the rise in the standard of pitch. Many have been repaired after accidents or have suffered from exposure to changing atmospheric conditions.

The first stage in the inquiry was to obtain the ' response curve '

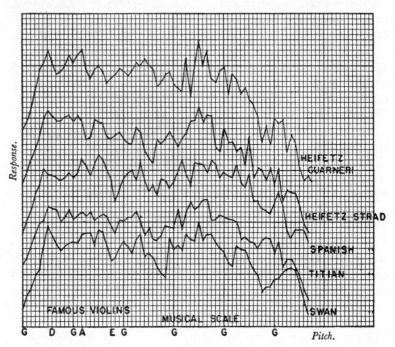

FIG. 7.10.—Response curves of famous violins. The first G is that of the open string

of each violin. This represents the intensity level (approximately the loudness) of each semitone given by the violin from the open G string to the E four octaves above the open E string. A method of bowing is used which eliminates the idiosyncrasies of the performer. In this way it was possible to get results which could be relied upon to repeat themselves. The curves for five famous violins are shown in Fig. 7.10. Each peak on these curves corresponds to a natural frequency of some part of the bridge–belly–back–contained-air system. All of them show a marked peak at C or C♯ on the G string. This is the resonance of the contained air—a fact which may be established by filling the violin with a dense gas, such as carbon dioxide, and repeating the experiment when the position of the peak shifts to a lower pitch. This resonance strengthens a range of pitch which would otherwise be weak. All the other prominent peaks represent some mode of vibration for the belly or back, as is indicated by the fact that they all shift towards the bass if the mute is mounted on the bridge.

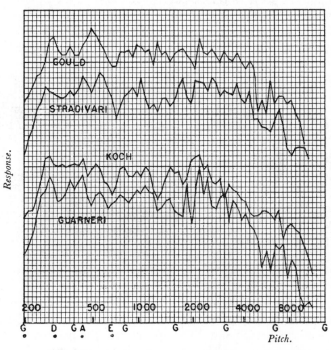

FIG. 7.11.—Response curves in pairs, one an old violin, the other a new violin

There is no agreement in detail between the curves, and they show that quality must change from note to note on a violin. To take an obvious case from the top curve, there is a peak at the second A and a depression near the third. Thus, if the second A is played, its second partial will be weak. On the other hand, there is a depression at the B just above it, and a peak near the octave above this, so that if the B is played, its second partial will be strong. The best violins seem to give a fairly uniform response level over most of the range, with a little excess in the region near the top of the E string (2,000–3,000 Hz) and a rather sharp dropping off to nothing at the highest pitches (5,000 to 10,000

Hz).   Thus the lower notes will be rich in partials and the highest notes relatively pure.   Fig. 7.11 shows for comparison two
pairs of response curves.   One of these curves in each case is for
a genuine old violin, the other for a modern copy, and it will
be seen that the old violin differs less from its copy than the old
violins differ among themselves.   It may be that with careful
modelling and good workmanship instruments as good as any of
those produced by the old craftsmen can still be made.   Saunders
is inclined to attach importance to the discovery that the work
required to make an old violin speak properly is notably less than

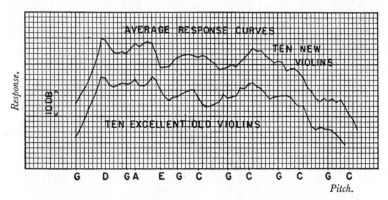

FIG. 7.12.—Average response curves for ten old violins and ten new ones

is required for a modern one.   This is measured by using a
mechanical bow and finding the critical pressure required to elicit
the tone of the instrument.   This means that the performer plays
with less effort and that the response of the instrument is much
quicker—a fact which is bound to tell in rapid passages.   As a
matter of fact, it is probable that the tone of the old violin is
something of a fetish.   This was tested before an audience listening
to a lecture on ' A Scientific Search for the Secret of Stradivarius '.
Three violins were played in rapid succession by an artist.   One
of them was a famous Stradivarius (the Rossignol, 1717), the
other two were very good modern instruments.   The result was
that about one-third of the audience got the Strad right, the
number which would have got it right by pure chance.   It is
admitted, however, that ' a few of the listeners, perhaps not more
than a dozen, could recognize the tone of a Stradivarius at once
and without hesitation ', although ' some persons who are justly

regarded as experts did not vote correctly '. This correct judgement may have been based on a recognition of the quickness of
response and economy of effort referred to above. Fig. 7.12
shows how much alike the tone-quality of old and new violins
must be when taken in groups. A string orchestra entirely
composed of Stradivarius instruments played in a concert in New
York in December 1937, and many listeners could hear nothing
unusual in their combined tone. This view is not pure philistin-
ism, as may be seen from the following quotation: [1] ' In the
opinion of many whose opinions are entitled to deference, the
master pieces of the greatest makers are gradually arriving at a
condition in which their value is appreciated by the curators of
museums rather than by great contemporary artistes . . . a fine
violin enjoys a sort of mysterious immortality, the effect of which
is often enhanced by the groundless idea that no good fiddles have
been made since the golden age of the Cremona makers, which
terminated 130 years ago, and that the secrets of violin making
are lost '.

The effect of varnish has been studied by Meinel in Berlin. It
is a difficult research to carry out. Different varnishes cannot be
tried on the same violin. We can only take the response curve
for a number of violins unvarnished and then test them again
after varnishing. Meinel found that the effect of the varnish was
to flatten out the peaks slightly, but not to shift them. The
magnitude of the effect was very small, and although it would
tend to improve performance, it seems doubtful if it could be
detected by ear. It is certainly no basis for the almost magical
effects attributed to Cremona varnishes.

A study has been made of vibrato tone by Seashore. Pitch
vibrato or tremolo appear in practically all tones of the violinists
studied. The average rate is 6·5 Hz, and the average extent
about 0·25 tone, which is only about half the average extent for
the singers measured. The rate and extent are maintained with
good consistency, and the form of the pitch pulsations is very
smooth and regular. Intensity vibrato is less common than pitch
vibrato, but when it appears it has about the same rate (6·3 Hz.)
and an amplitude of 2·2 phons.

Plucked String.—This has played an important part in music in
the past through its use in the harpsichord, spinet, virginal, lute,
lyre, and psaltery. The only important orchestral instrument is

---

[1] Grove's *Dictionary of Music and Musicians*, 1911 Edition, Vol. V, p. 287.

the harp.   The name and the modern form of the instrument come from Northern Europe, the earliest pattern being depicted in manuscripts of the ninth century.   It was very greatly improved by Sebastian Erard in 1810, when he perfected a pedal mechanism which shortened the strings so as to raise the pitch either by one or by two semitones.   There are seven pedals, each acting on one note, and altering the pitch of that note in every octave throughout the range.   The harp is tuned in the key of C♭.   Depressing all the pedals to the first notch gives us the key of C natural.   Depressing the F pedal a further notch gives F♯, and hence the key of G natural.   The mechanism is ingenious, and is well concealed in the pedestal, the vertical pillar, the neck, and the comb, all of which are more or less ornamental in design.   There are forty-six strings, varying in length and tension, and giving a compass of $6\frac{1}{2}$ octaves.   Those in the upper and middle register are of gut, while those in the bass are of covered steel wire.

The distribution of partials for a plucked string follows the same general principles as in the case of the struck string or the bowed string.   It depends on

   (a)   the point of attack,
   (b)   the instrument of attack.

(a)   It is broadly true that a partial is absent or nearly absent if the string is plucked where a node is required, and is a maximum when the plucking is at the position of an antinode.   It is also true, however, that given the favourable position for eliciting a partial, the possible intensity of the higher partials falls off very rapidly.   The dissonant partials are eliminated, as in other stringed instruments, by applying the fingers at about $\frac{1}{7}$th to $\frac{1}{9}$th of the length of the strings from the end.

(b)   The quality is greatly influenced by the choice of the instrument of attack.   In the harp the fingers are used, the ' instrument ' being broad and soft.   This avoids sharp angles in the displaced string and tends to diminish the importance of the highest partials, giving a sweet and mellow quality.   On the other hand, in the mandoline, where a hard plectrum is used, the string or wire is pulled out to a sharp angle, and the highest partials are relatively strong, giving a metallic quality to the tone.

The motion of a string when plucked and released is quite different from the motion of the struck or bowed string.   Let ACB (Fig. 7.13) represent the string with the point C pulled aside ready to let go, the scale being greatly exaggerated. Complete the parallelogram ACBD. Then the successive positions of the string during the vibration are ACB, AEFB, AGB, AHKB,

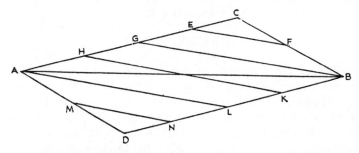

FIG. 7.13.—Motion of plucked strings. The motion is given by assuming the line HK to move to and fro between C and D, remaining parallel to itself. Successive shapes are ACB, AEFB, AGB, AHKB, ALB, AMNB, ADB

ALB, AMNB, ADB, and back through the same intervening steps to ACB. Thus the successive shapes of the string are as shown in Fig. 7.13. The displacement diagram for any point of the string is shown in Fig. 7.14. The point moves with uniform

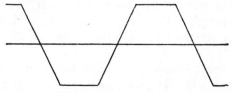

FIG. 7.14.—Vibration curve of a point on the string

speed to its position of maximum displacement, then stops for a time, returns with uniform speed to maximum displacement on the other side, and then stops for a time again.

CHAPTER 8

ORGAN PIPES

Vibrations of Air Columns.—If we think of an air column in a tube open at both ends, and try to imagine the ways in which it can vibrate, we shall readily appreciate the fact that the ends will always be antinodes, since here the air is free to move. Between the antinodes there must be at least one node, and here the air is preserved at rest because, at the ends, the moving air is either moving towards the centre from both ends or away from the centre at both ends. Thus the simplest kind of vibration has a node at the centre and antinodes at the two ends, the vibration being represented in Fig. 8.1a, where the arrows above represent the velocity of the air in magnitude and direction during one-half of the vibration, while those below represent it during the other half of the vibration. The wave-length λ_1 of the vibration is four times the distance from node to antinode $= 2l$, where l is the length of the pipe. This, then, is the wave-length of the simplest method of vibration—that of the fundamental of the pipe.

To get the next simplest we must insert another node and another antinode in the pipe, and this will give the second partial. The method of vibration is shown in Fig. 8.1b. The wave-length λ_2 is now l. Thus the wave-length of the second partial is half the wave-length of the first partial—i.e., it is the octave.

To obtain the third partial we insert another node and another antinode in the length of the pipe, and obtain the mode of vibration shown in Fig. 8.1c. This time the wave-length $\lambda_3 = \dfrac{2l}{3}$, which is one-third of the wave-length for the fundamental mode. Since the frequencies are inversely proportional to the wave-lengths, we see that the modes of vibration of a column of air in a pipe open at both ends give the full series of harmonic partials.

Let us now consider what happens if the pipe is closed at one end. In this case one end must always be a node, and the other an antinode. In the simplest mode of vibration the movement of the air will be as shown in Fig. 8.1d. The air at the open end will move in and out alternately, the amplitude of the vibration being greatest at the opening end and diminishing as we approach the closed end. The distance from node to antinode

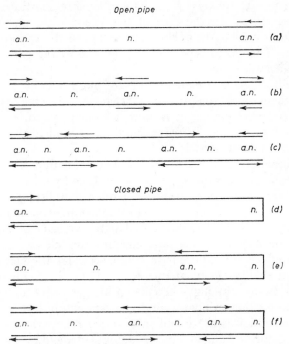

FIG. 8.1.—The first three modes of vibration of the air in an open pipe and in a closed pipe

is in this case l, the whole length of the pipe. The wave-length λ'_1 is therefore $4l$.

To get the next simplest mode of vibration we insert another node and antinode, as shown in Fig. 8.1e. This gives us a wave-length $\lambda'_2 = \dfrac{4l}{3}$.

The third partial is shown in Fig. 8.1f. Here we have a wave-length $\lambda'_3 = \dfrac{4l}{5}$.

When we compare in this way the possible modes of vibration of a pipe open at both ends with a pipe closed at one end, two important differences appear:

(1) The fundamental tone for the open pipe has a wave-length $2l$, while that for the fundamental of the closed pipe is $4l$. Obviously, then, the open pipe gives the octave of the closed pipe of the same length.

(2) The wave-lengths for the partials of the open pipe are in the ratio $1/1$, $1/2$, $1/3$, &c. They therefore constitute the full harmonic series. In the case of the closed pipe, on the other hand, the wave-lengths of the partials are in the ratio of 1, $1/3$, $1/5$, &c. The closed pipe therefore gives only the odd members of the harmonic series.

The vibrations in a conical pipe are more difficult to forecast, but mathematical analysis shows that in this case the closed conical pipe behaves as if it were an open cylindrical pipe of the same length—i.e., its wave-length is $2l$, and it gives the full harmonic series of partials.

The situation is much more complex than that simple result might lead one to suppose, however. It is found on closer inspection that while the antinodes remain equidistant as for the cylindrical pipe, the nodes are all shifted, some considerably and others slightly.[1] It is found also that the velocity of sound is a function of frequency in a tapering pipe, the change in velocity being in effect just sufficient to compensate for the fact that the pipe is in reality closed at one end. Thus the frequencies the pipe can sound, although they are in integer ratios and *appear* to be the same as for an open cylindrical pipe, do in fact result from the same modes as are found in a closed pipe; the frequencies of each mode is ' pulled ' by an amount depending on the frequency. For example, the frequencies of the first four modes of a tapered pipe closed at one end should be in the ratios $1:3:5:7$; however they turn out to be in the ratios: $1:2:3:4$. The frequency of the second mode has been ' pulled ' from being three times that of the fundamental to being twice the fundamental frequency, and so on. This point is discussed at some length in the literature.[2,3,4,5] A further paper that is of relevance here is that by Lange.[6]

These results can be roughly verified by taking three card-

[1] Barton, *Phil. Mag.*, Vol. 15, p. 69 (1908).
[2] Long, *J. Acous. Soc. Amer.*, Vol. 19, p. 892 (1947).
[3] Webster, *ibid.*, Vol. 19, p. 902 (1947).
[4] Young, *ibid.*, Vol. 20, p. 345 (1948).
[5] Long, *ibid.*, Vol. 20, p. 875 (1948).
[6] Lange, *Acustica*, Vol. 5, p. 323 (1955).

board tubes, (*a*) a cylinder open at both ends, (*b*) a cylinder closed at one end, (*c*) a cone closed at the apex, and tuning them all by adjusting the length so that they reinforce a fork of frequency 256. The relative dimensions are shown in Fig. 8.2. If now they are all three tested by holding over them in turn a fork of frequency 512, it will be found that (*a*) and (*c*) respond, but not (*b*). On the other hand, all three will respond to a fork of frequency 768, showing that all can vibrate to give the frequency 3 × 256, although the closed cylindrical pipe is incapable of giving a frequency 2 × 256.

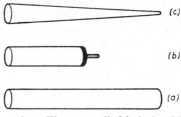

FIG. 8.2.—The open cylindrical pipe (*a*) and the closed conical pipe (*c*) are of the same length. The closed cylindrical pipe is of half the length. All three give the same note.

The actual motion of the air in a tube that is sounding a partial tone can be studied in several ways, two of which may be mentioned here.

(1) A very small coil of wire, heated by an electric current, may be fixed to a thin rod and pushed along the pipe. Where there is an antinode, the air will be in rapid motion and the hot wire will be slightly cooled. This cooling lowers its electrical resistance, and the change can be detected at once. At a node, on the other hand, the air is at rest, and no such cooling takes place. The hot wire thus enables us to locate the antinode.

(2) An old experiment due to Kundt (1839–1894) used lycopodium powder—a very light dust—scattered in a horizontal tube to reveal the motion of the air when it was in vibration. The dust gathered in equidistant stagnant heaps at the nodes of the motion. Fig. 8.3 shows a modern variant of the same experiment. A horizontal tube is closed at one end by an adjustable piston and at the other end by a telephone diaphragm which can be made to vibrate at a controlled frequency. Lycopodium powder or fine cork dust is sprinkled along the bottom of the tube. If the frequency is adjusted to a partial of the air column, the dust forms little ridges at the antinodes, where the air is in motion, like the ridges impressed on the sand by wind, and very similar in cause. At the nodes the dust is at rest. If the motion of the air is sufficiently violent, the dust

rises up at the antinode and forms a disc right across the tube. These discs locate the antinodes very exactly. If the tube is filled with tobacco smoke and a strong beam of light passed down the tube, a microscope can be focused on a smoke-particle which, when the air is at rest, will be seen as a slowly moving, bright speck. When the tube is made to sound, the air is at once thrown into vibration, and if the smoke-particle is small enough, it exactly follows the motion of the air, vibrating very rapidly in a line parallel to the axis of the tube. It will therefore appear as a bright horizontal line, the length of which indicates the amplitude of the vibration of the air at that point in the tube. By examining smoke-particles at different points, we can see how the amplitude of vibration varies, and can thus locate the nodes and antinodes.

It will be noticed that all the frequencies depend on c, the velocity of sound, and as the velocity of sound depends on temperature, the pitch of a vibrating air column will also vary with the temperature, a phenomenon which has already been discussed on p. 49.

Aeolian Tones.—Long before man could produce musical sounds, the wind was producing them. These sounds varied continuously in pitch and were produced in all sorts of different ways. Before we consider the most interesting of them, let us take a look at some other phenomena which, although they seem quite remote from the sounds produced by wind, are really very closely related. Try to draw your hand quickly through your bath-water with your fingers loosely spread, and you will find that your fingers tend to strike against one another, vibrating from side to side across the track of your hand. Repeat the experiment with the handle of a brush, and you will notice the same tendency. Watch the surface of the water carefully in light reflected from the electric lamp, and you will see that as the handle is drawn forward it leaves in its train a procession of little dimples on the water surface, these being formed alternately from the two sides of the brush. The dimples are little whirl-pools caused by the motion of the brush-handle. Sometimes their shadow can be seen on the bottom of the bath. What is really happening is the formation of a series of those little whirl-pools, spaced as shown in Fig. 8.4, Plate IX, and having obviously been shed from opposite sides of the obstacle. The same effect would be produced by holding the handle of the brush steady in flowing water. Each little whirlpool or vortex urges the handle to the side as it is shed and, if the water is moving with uniform speed, the handle is urged to and fro across the stream with uniform frequency. Anyone who has fished with a long line in a strong tide will remember the throbbing of the line. Sometimes the anchor-rope of a small boat gives the effect when the boat is anchored in deep water for fishing and the tide runs strongly. The frequency of the motion of the obstacle—handle, line, rope—depends only on two things: the diameter of the obstacle and the velocity of the stream.

Now, what is true of water flowing past an obstacle, or of an obstacle drawn through water, is equally true if we substitute air for water. In fact, it is a property of all fluids, gases and liquids alike. The fluttering of a flag in the breeze is due to the flag-pole acting as an obstacle to the air-stream. Eddies are shed from opposite sides of the pole alternately, and chase one another along opposite sides of the flag. It is found that the frequency of the formation of eddies, and hence the frequency of the vibration imposed on the obstacle and of the resultant note, is given approximately by

$$f = \frac{V}{5d}$$

where $\left.\begin{array}{l} V = \text{velocity of air-stream} \\ d = \text{diameter of obstacles} \end{array}\right\}$ in compatible units.

When there is a wind blowing we often notice the singing of telegraph wires—a singing which communicates itself from the wires to the posts and can be well heard when we press an ear to the post. The response of the wire to the action of the wind will, of course, be strongest if the wire on which the wind acts is stretched so that its natural frequency of vibration is the same as the frequency imposed on it by the wind. The effect of the diameter of the obstacle on the frequency is quite obvious if we compare the roaring of the wind in a forest (large diameter of branches, low frequency of note) with the whistling of wind over long grass on the uplands (small diameter of grass stems, high frequency of note). The effect of the velocity of the air past the obstacle can be observed if, instead of letting air stream past the obstacle, we drive the obstacle through the air. If, for instance, we take a thin cane and 'swish' it through the air, we shall have no difficulty in observing that the more rapidly we move it the higher is the pitch of the note which it gives.

Various references occur in early literature to the ' Aeolian Harp ', an instrument designed to use this phenomenon for musical effect. The first systematic account is said to occur in a seventeenth-century book by Athanasius Kircher. Later instruments were made with wires of different thicknesses tuned in unison.

Edge Tones.—If air is forced through a slit, the conditions for the formation of eddies are again present, and these eddies are formed as shown in Fig. 8.5, alternately on opposite sides, the jet itself taking the form there shown. Here again a note is produced, but it is very feeble and indecisive, and quite unlike the Aeolian tone which we have just been discussing. Its frequency depends on the velocity of the air and the width of the slit. It is known as a ' slit tone '. About the middle of the nineteenth century, however, it was noticed that if an ' edge '— a wedge of very small angle—was presented centrally to the jet coming from the slit, the note was stronger, steadier, and more reliable. A very complete study

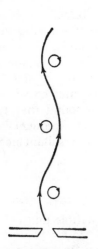

FIG. 8.5.—Jet from a slit showing sinuous form and vortices

of the phenomenon has been made by Brown.[1] He agrees with earlier experimenters that the frequency of the edge tone depends on (*a*) the velocity of efflux of the air from the jet and (*b*) the distance of the edge from the slit. The relationship is not very simple. If the edge is first of all placed quite close to the slit, no sound is heard. If the edge is now slowly withdrawn, a point is reached at which a clear, bright tone is heard. As the edge is further withdrawn, the pitch falls steadily until it suddenly makes a considerable jump upwards in pitch. Once more the pitch falls with increasing distance between edge and slit until a second jump occurs. Later a third jump takes place, and after this at a certain distance the note ends in an irregular and confused noise. Fig. 8.6, Plate IX, shows photographs of the four stages, and they seem to be connected with the number of bends shown by

[1] *Proc. Phys. Soc.*, Vol. 49, p. 493 (1937).

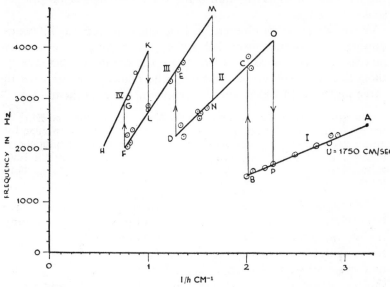

FIG. 8.7.—Variation of frequency of edge tone with reciprocal of wedge-distance

Stage.	Diagram.	Frequency change.
I	AB BC	Gradual fall to 1,500 while h increases to 0·5 cm. Sudden jump to 3,800
II	CD DE	Gradual fall to 2,300 while h increases to 0·77 cm. Sudden jump to 3,500
III	EF FG	Gradual fall to 2,000 while h increases to 1·33 cm. Sudden jump to 2,900
IV	GH	Gradual fall to 2,000 while h increases to 1·8 cm. Here the note disappears.
IV	HK KL	Gradual rise to 4,000 while h diminishes to 1 cm. Sudden drop to 2,800
III	LM MN	Gradual rise to 4,800 while h diminishes to 0·6 cm. Sudden drop to 3,000
II	NO OP	Gradual rise to 4,100 while h diminishes to 0·43 cm. Sudden drop to 1,700
I	PA	Gradual rise to 2,500 while h diminishes to 0·31 cm. Below this value of h the note ceases.

the jet, and therefore the number of eddies formed before the jet strikes the edge.

A similar series of changes in frequency is observed if, maintaining the distance from slit to edge constant, we gradually increase the air velocity. In this case the frequency increases as the velocity of the air is increased, the change taking place in three jumps, with steady rises in frequency between them.

The graph in Fig. 8.7 gives the relationship between the frequency of the edge tone and the reciprocal of the distance from slit to edge for an air velocity of 1,750 cm./sec. We may describe what happens as follows: the edge is drawn gradually away from the slit, and no note is heard until $h = 0.31$ cm., when a note of frequency 2,500 starts to sound. This is the point A ($1/h = 3.2$ cm.$^{-1}$).

Flue Organ Pipes.—Much of our music depends on the vibrations of air columns. Almost all our wind instruments, including the organ, consist of one or more air columns set in vibration in one of their normal modes and maintained in vibration for a longer or shorter period by an air-stream under pressure. Of these columns the simplest in construction is the flue-pipe. It is also the oldest, and it forms the basis of the tone-production in the organ. When made of metal it is circular in section, and its main features are shown in Fig. 8.8, which represents an open diapason pipe. The air enters at the tip A and passes through the foot B, a conical portion of the pipe, to the flue C, a narrow slit which gives its name to the pipe. The tip can be closed up or opened out easily by working with tools, and in this way the supply of air to the pipe may be adjusted. The flue is bounded on the outer side by the lower lip

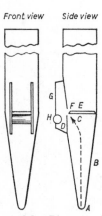

Front view   Side view

FIG. 8.8.—Diagram of flue-pipe showing: tip, A; foot, B; flue, C; lower lip, D; languid, E; upper lip, F; ears, G; beard, H

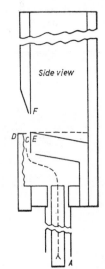

Side view

FIG. 8.9.—Section of stopped diapason organ pipe showing: tip, A; flue, C; cap, D; languid, E; upper lip, F

D and on the inner side by the languid E. The flue directs
the air-stream on to the upper lip F, and the distance from
the flue to the upper lip is known as the 'cut up', and is
of the greatest importance in determining the volume and
quality of the tone of the pipe. The ears G project at right
angles on each side of the pipe, but are not a constant feature.

The wooden flue-pipe is illustrated by the stopped diapason,
shown in Fig. 8.9. The tip A plays the same part as in the
metal pipe, but the wind supply is adjusted by inserting small
pieces of wood. The languid E is made of hard wood, usually
mahogany, and is shaped so as to give the desired quality by
altering the form of the wind-way. The cap D forms the flue C
with the block, and directs the wind on to the lip F.

Stopped flue-pipes may be made either of wood or of metal.
Wooden pipes are rectangular in section, and the stop is a rectangular block of wood covered with leather and fitted with a
handle. Metal pipes may be closed with a stop of wood or
metal covered with leather, or by a cap fitted air-tight over the
end of the pipe.

Obviously we have here the conditions necessary for the production (a) of an edge tone as the sheet of air from the flue
impinges on the upper lip, (b) of the note due to a vibrating air
column as the air in the body of the pipe is set in vibration by
the edge tone. These two vibrations, in fact, constitute a coupled
system, with the air column very much the predominant partner
in the concern owing to its relatively great mass. The system
will therefore vibrate nearly with the frequency of one of the
natural modes of vibration of the air in the pipe, selecting the
mode which gives a frequency nearest to the natural frequency
of the edge tone. The frequency of the edge tone depends, of
course, on the speed of the air (i.e., on the blowing pressure)
and on the height of the cut up. If, then, the blowing pressure
is fixed, the height of the cut up must be so chosen that the
natural frequency of the edge tone is also the natural frequency
of the vibrating column. In this case neither system tends to
force the vibrations of the other, and both vibrate in their own
natural frequencies.

That this is the true explanation of the action of the pipe is
borne out by the phenomena of 'over-blowing' and 'under-
blowing'. As the wind-pressure is increased the pitch of the

edge tone tends to rise, and so to force up the pitch of the air in the pipe. This responds very little, however, until the pitch of the edge tone is nearer to that of the second partial of the pipe than to the fundamental, when the note of the pipe jumps and, if it is an open pipe, gives the octave. As a matter of fact, this statement is a slight oversimplification, as will be seen from Fig. 8.10, which is taken from a paper by Mokhtar.[1] The continuous lines give the frequency of the note of the coupled

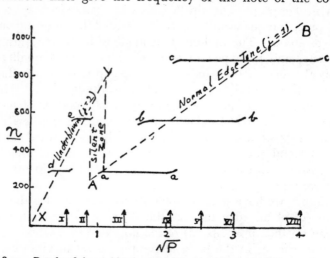

FIG. 8.10.—Result of 'over-blowing' and 'under-blowing'. The continuous line shows the tone of the pipe. I, II, III, &c., refer to the stages in Fig. 8.11, Plate X. P is measured in centimetres of water

system. The dotted lines show the natural frequencies of the edge tone plotted against the square root of the wind-pressure P. Starting at a with $\sqrt{P} = 1 \cdot 2$, we see that the pipe gives its fundamental. As the wind-pressure is increased the natural frequency of the edge tone rises, but it is held by the vibration of the pipe, and there is no appreciable change of pitch until at $\sqrt{P} = 1 \cdot 5$ the octave comes in at b, and octave and fundamental sound together, the octave at first being distinctly flat. As $\sqrt{P}$ begins to exceed 2 the edge tone pulls the fundamental sharp, and the fundamental disappears, its place being taken by the twelfth, which comes in at c almost, if not quite, before the fundamental has disappeared.

A similar effect is elicited by under-blowing. The pitch of the

[1] *University of Durham Philosophical Society*, Vol. 9, p. 352 (1938).

coupled system must be one of the natural frequencies of the
pipe. But at low pressures the normal frequency of the edge
tone is lower than the fundamental frequency of the pipe.
There are other possible frequencies for the edge tone, however,
corresponding to the other stages previously discussed. As soon
as one of these lies near the fundamental of the pipe, the note
begins, and as the air-pressure is increased it jumps to the octave,
after which there is a silent zone, and the edge tone jumps to its
normal mode and the pipe gives its proper fundamental. This
sounding of the octave when the pipe is under-blown is a frequent
defect, which has earned for one type of organ pipe the name of
the ' coughing ' bourdon. With a stopped pipe similar results
are obtained, except that when the pitch of the pipe jumps it
goes straight to the twelfth, the stopped pipe not giving even
partials. Fig. 8.11, Plate X, shows the oscillograms for the
vibrations of the air. I shows the under-blown fundamental at
$\sqrt{P} = 0.26$. In II at 0.65 the fundamental has disappeared and
the under-blown octave has taken its place. III shows the co-
existence of the fundamental with other partials. IV, V, VI, VII
give the form of the vibration at the pressures marked in Fig. 8.10.
VIII, IX, and X are oscillograms for $\sqrt{P}$ 1.46, 6.00, and 16.00,
respectively, and indicate three stages in the production of over-
tones of a stopped pipe for which the fundamental (shown on
VIII) had a frequency of 230 Hz. X is obviously the second
partial for a closed pipe—i.e., the twelfth.

Reed Organ Pipes.—Flue-pipes form the foundation of organ
tone, but reed-pipes are an important part of the superstructure,
owing to their distinctive quality of tone. The construction of
the pipe is shown in Fig. 8.12. The reed itself is a thin strip of
brass fixed at the upper end by a wedge and vibrating against
the triangular aperture in a shallot. In this case it is a ' beating ' reed, striking against the face of the shallot at each vibration. ' Free ' reeds are sometimes used, which can just pass
through the aperture over which they are fixed. The reed itself
may be loaded with a small weight attached to its lower end.
The shallot fits into a lead block, as does also the shank of the
pipe. A wire tuning-spring is braced against the reed, and may
be used to lengthen or shorten the vibrating length of the reed.
The reed and shallot are enclosed in a ' boot ', generally made of
pipe-metal, the tip of which carries the hole through which the
wind enters. The reed-pipe may be looked on as a coupled

system similar to the flue-pipe, except that the edge tone is replaced by the reed, but the phenomenon is really a little more complicated than this.   The air in the pipe, the air in the boot, and the reed are all coupled together.   The air contained in a long boot may have a frequency of the same order as the pipe and reed, and may ruin the resultant tone.   To prevent it vibrating in its fundamental mode a hole is sometimes pierced in the boot.   Alternatively a very short boot may be used, and this has the advantage that the pipe speaks more promptly.

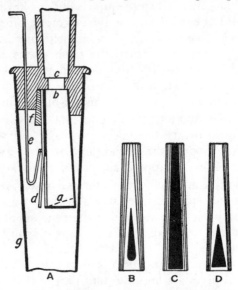

FIG. 8.12.—Reed-pipe. A, beating reed; B, filled-in shallot; C, open shallot; D, closed. *a*, head of shallot; *b*, tip of shallot; *c*, tip of tube; *d*, tongue; *e*, tuning spring; *f*, wedge; *g*, boot

Examination of the motion of the reed shows that the reed opens and air enters at the moment when the pressure in the pipe is greatest.   This is the opposite of what takes place at the open end of a pipe, air in this case beginning to enter at the instant when the pressure is a minimum.   The reed therefore acts as a ' closed end ', and we find that whereas the end at which air enters a flue-pipe is an open end, that at which air enters a reed-pipe is, in this sense, a closed end.   We should therefore expect an open reed-pipe—i.e., one open at the upper end—to act like a flue-pipe closed at the upper end.   This we find to be the case.

In the vibrating partnership the pipe is the predominant

partner, as in the case of the flue-pipe, but the reed tone has more influence on the resultant pitch than the edge tone, and the pipe is tuned by tuning the reed. The pipe or body is usually designed so that the natural frequency of the contained air is the same as the natural frequency of the reed, but this is not always the case (see p. 128).

Flue-pipes are made rectangular in section if of wood, or cylindrical if of metal. Reed-pipes are almost always conical, with the exception of the clarinet and the vox humana, which are cylindrical. The oboe is a slender, tapering cone terminating in a ' bell ' or cone of wider flare.

Tuning of Organ Pipes.—Flue-pipes are tuned by adjustment of the pipe, and reed-pipes by adjustment of the reed. Generally speaking, any change which impedes the motion of the air at the open end of a pipe is equivalent to lengthening the pipe, and lowers the pitch, while any change which facilitates the motion of the air is equivalent to shortening the pipe, and raises the pitch. Thus open flue-pipes are frequently fitted with a tuning device. This may be a slot close to the end cut parallel to the length of the pipe and partially closed by an adjustable sliding piece known as a tuning-piece or a regulating slide. Opening a larger area of the slot sharpens the pitch, and vice versa. Flue-pipes can also be tuned by a direct alteration of the effective length of the pipe. This is effected by a kind of sleeve, which fits over the end of the pipe and can be moved up or down. It is known as a clip, and is applicable, of course, only to pipes of cylindrical or rectangular form, and not to conical pipes. Metal pipes without tuning devices can be tuned by the use of cones. These tuning-cones are either solid or hollow. In the former case the apex of the cone is inserted into the pipe, and the base of the cone is gently tapped. In this way the mouth of the pipe is slightly widened and the pitch raised. In the case of the hollow cone or ' cup ' it is placed over the end of the pipe, and the vertex is tapped. This has the effect of slightly closing the end of the pipe, and so flattening the pitch. Stopped pipes are tuned by adjusting the stopper, which in the case of wooden pipes is a rectangular piece of wood covered in leather and fitted with a handle.

Reed-pipes are tuned by adjusting the tuning-spring by gently tapping it up or down, so as to lengthen or shorten the vibrating length. Reed-pipes are sometimes fitted with clips or slots, but these are for the adjustment of the pipe to the reed, and not for tuning.

Quality of Tone in Organ Pipes.—Organ tone is infinitely variable, but there are four main categories generally recognized:

(*a*) Flute tone.   (*c*) String tone.
(*b*) Diapason tone.   (*d*) Reed tone.

Of these, the first three are produced by flue-pipes, and the last, as the name suggests, by reeds. Diapason tone is regarded as the basis and foundation of organ tone. From the point of view of quality it comes between the purity of the flute tone and the relative richness of string tone, but Bonavia-Hunt says of it, ' It does not bind flute to string, as is commonly supposed, nor is it merely the substratum of organ tones. It *is* organ tone, normally, fundamentally, characteristically.' There are, of course, no physical standards as yet which can be used to define these various categories of tone. That will come—is, indeed, already on the way. But so far we are dependent on tradition and on aesthetic judgement, and the modifications of a pipe to give some characteristic quality of tone is a matter of judgement.

Considering first the flue-pipe, the following variables are at our disposal, and are related to quality in the ways indicated according to fairly general agreement among experts.

(1) *Material of pipe.*—It has sometimes been held that the sound produced by an air-cavity such as an organ pipe depends only on the geometric form of the cavity, and not at all on the material forming the walls of the cavity. This view has never been popular among musicians, and has now been universally abandoned. The material of which a pipe is made does affect the quality and, for pipes identical in shape and size, the pitch. In general terms it is agreed that there is justification for the views of organ-builders who claim that heavy lead alloys emphasize the lower partial tones, whereas with zinc the higher partials are amplified and the quality of tone is brighter and more strident. The results of attempts so far made to confirm this view by physical analysis show clearly that the material used does affect the quality, although the measurements [1] indicate less decisive differences than might have been anticipated.[2] Of eight

[1] Boner and Newman, *Journ. Acoust. Soc. Amer.*, Vol. 12, p. 83 (1940).

[2] For a contradictory view see Knauss and Yeager, *Journ. Acous. Soc. Amer.*, Vol. 13, p. 160 (1941), who maintain that the walls have no effect on the tone quality of the cornet and describe several experiments which confirm their view.

materials tried, common pipe metal gave the best reinforcement of the first seven partials, with wood second, galvanized iron third, and a heavy copper pipe fourth. Over the partials from eight to eleven steel gave the best reinforcement, with wood second, the pipe metal having dropped to seventh place. For partials above the eleventh, the order was much the same. Steel was still first, with a medium-weight copper pipe second, while the wood had dropped to fourth place, and the pipe metal remained at seventh place. The fact that steel, which is so good in some respects, has never won a place for itself in organ-pipe construction is due, no doubt, to the fact that its tone is all ' tops ' and no foundations. It only comes fifth in the order for the first seven partials. It is the superiority of pipe metal and wood over this range which probably accounts for the secure place of these materials.

Pipe metal is an alloy of lead and tin, and use prescribes different compositions for different qualities. Thus for flute and diapason tone 30 per cent. of tin is common, while for string tone 45 to 55 per cent. of tin is used. Alloys with 35 to 65 per cent. of tin cool with a spotted surface of striking appearance and the alloy is known as ' spotted metal '.

(2) *Wind-pressure.*—This is an important factor, and must be considered not only from the point of view of the pressure in the bellows, but from the point of view of the pressure at the flue (or reed in the case of reed-pipes). This is regulated by paring or cupping the aperture through which the wind is admitted to the pipe, if of metal. The cupping process is similar to that used for flattening the pitch of a pipe by diminishing the aperture of the open end. In the case of wooden pipes, regulation requires the insertion or abstraction of wooden plugs. The higher the pressure of the wind the brighter is the resulting tone and the higher its proportion of partial tones. The pressure is regulated for the right quality of tone by ear. In the old organs the pressure was as low as $2\frac{3}{4}$ inches of water.[1] In modern organs 3 inches is about the lowest in common use, reed-stops may use from $5\frac{1}{2}$ to 8 inches, while tubas (a large-scale reed-pipe) may use pressures as high as 20 to 25 inches.

(3) *Scale.*—The scale of a cylindrical pipe is its internal diameter in inches. In the case of a wooden pipe the width of the mouth is sufficient to indicate the scale, although for

[1] Organ pipe wind pressures are traditionally measured in inches, even on the continent.

exact statement the two internal dimensions would be required. The scale for the lowest pipe in a stop fixes the scale for all, as they are graduated according to a fixed relationship. For instance, in the case of large-scale pipes they may reduce to half at the seventeenth note of the scale. In the case of small-scale pipes this rate of reduction might be too great, making the upper pipes too small.

The influence of the scale of the pipe on the quality of the tone is in part due to what is known as the ' end correction ' of a pipe. In considering the vibrations of the air in a pipe we assumed that the open end would be an antinode or place of maximum motion. This is approximately the case, but in point of fact the antinode is situated just beyond the end of the pipe. Thus for a pipe open at both ends the half wave-length of the vibration (the distance between two antinodes) is a little greater than the length of the pipe, and the frequency of the vibration is therefore a little less than would be calculated from the simple formula given on p. 110. For a cylindrical pipe of radius r with a sharp and clean end the distance from the end of the pipe to the antinode is $0 \cdot 6r$. Any obstacle shading the end, any bending of the edge inwards (as in tuning), obstructs the motion of the air, throws the antinode farther from the end of the pipe, increases the wave-length of the sound, and so flattens the pitch of the pipe. On the other hand, anything that facilitates the vibration of the air, such as bending the edges outwards or opening a slot near the end, brings the antinode closer, diminishes the wave-length, and sharpens the pitch. At the flue end of the pipe the obstruction is considerable, and the correction to be added to the true length of the pipe to give its effective length is about $2 \cdot 7r$. For an open flue pipe whose measured length is l the half wave-length is $l + 0 \cdot 6r + 2 \cdot 7r = l + 3 \cdot 3r$. This accounts for the fact that a pipe always gives a note of lower pitch than its measured length, uncorrected, would lead us to expect. An open diapason pipe may have a length of 8 feet and a radius of 3 inches. The ratio of the corrected length to the uncorrected length is $\dfrac{96 + 9 \cdot 9}{96}$. This is the reciprocal of the ratio of the frequencies, which is therefore $\dfrac{96}{106}$ approximately, an interval of about a tone.

If the correction were the same for all the partial tones of the pipe, then these partial tones would all be in tune with the harmonics of the edge tone, since their frequencies would be in the ratio $1:2:3:4$, &c. This, however, is found not to be the case. The correction is different for the partial tones, which are therefore out of tune with the harmonics of the edge tone, and so are diminished in intensity. The greater the radius of the pipe the greater is the correction, and therefore the greater the mistuning, so that we should expect narrow-scale pipes to be richer in high partials than wide-scale pipes. Further, it is found [1] that the highest partial which can be evoked from a pipe is one whose wave-length is four times the diameter of the pipe. Therefore the narrower the pipe the shorter the wave-length, and the higher the frequency of the highest partial which can be generated.

These results agree with the judgement of organ-builders and others. String-toned pipes—viole d'orchestre, viola da gamba, &c.—are all made of very small diameter, while flute stops are all made of wide diameter and give relatively pure tones.

(4) *Cut-up.*—This is the height from the flue to the upper lip. It is sometimes stated as a fraction of the width of the mouth— e.g., a cut-up of one-fourth. We have seen, in discussing the edge tone, that its pitch rises with the wind-pressure and falls as the cut-up is increased. It follows that to get the edge tone of the right pitch cut-up and wind-pressure must be mutually adjusted. A very low cut-up gives string tone, a higher one gives diapason tone, while a higher one still gives flute quality. If the cut-up is very high, the tone becomes practically pure. On the other hand, a high cut-up can be compensated by high wind-pressure, and this gives added power.

(5) *Width of mouth.*—In wooden pipes the mouth is cut the entire width of the body. In metal pipes it is a definite fraction of the circumference—e.g., one-third. A wide mouth is said to favour power and ground-tone, and a narrow mouth to favour upper partials. Thus diapason tone involves a wide mouth, and string tone a narrow one.

(6) *Nicking.*—This process is carried out on the languid and bottom lip of metal flue-pipes. A small, spear-pointed instrument is used to produce V-shaped depressions said to steady the tone and eliminate a snarling quality. In the case of wooden pipes the nicking consists in making file-cuts at an angle to the grain on the block alone, or on the cap alone, or on both. In

[1] Anderson and Osten, *Phys. Rev.*, Vol. 31, p. 267 (1928).

the case of diapasons, nicking which is deep but not too close is recommended, while for gambas and small-scale pipes it is suggested that the nicking should be shallow and close. Recently [1] the use of nicking, at least in the case of the larger pipes, is discouraged.

(7) *Treatment of upper lip.*—If this is blunt the resulting tones are nearly pure, while if the lip is sharp a cutting quality of tone is generated. In the case of diapasons the upper lip is sometimes covered with leather, to give mellowness of tone and improve speech.

(8) *Roller beard.*—The beard or bridge is a piece of wood of circular or oval cross-section held between the ears of the pipe just in front of the mouth (Fig. 8.8). Its effect is to enable a higher wind-pressure to be used without over-blowing the pipe. It is used principally with gambas, which are first over-blown to the octave, and then the beard adjusted until the fundamental tone is resumed.

What has been said under (1), (2), and (3) above on material, on wind-pressure, and on scale, applies to reed-pipes as well as to flue-pipes. But much the most important factor in modifying the tone of a reed-pipe is the reed. The striking reed may produce a sudden and complete stoppage of the air-stream in each vibration, and this produces an air vibration very rich in partials. Indeed, the resulting tone has been described as ' fiery '. If, however, the reed is ' voiced ', it is given a curvature, as a result of which it does not close the aperture in the shallot suddenly and completely, but rolls down over the aperture, cutting off the air-stream more gradually. This voicing of the reed is carried out by a steel instrument known as a ' burnisher ', and giving the tongue the correct curvature is a highly skilled job, requiring a large degree of musical taste.

Both flue-pipes and reed-pipes may be harmonic—i.e., they may sound the second or third partial of the fundamental of the pipe. The ' harmonic flute ', for instance, has a length corresponding to a note an octave below its actual speaking note. The second partial is elicited by perforating the pipe with a hole of the required size near the middle, and so encouraging the formation of an antinode there. The ' Zauberflöte ' or ' harmonic gedeckt ' is a stopped harmonic flute perforated at one-third of its length from the closed end, and so giving the twelfth of its fundamental. In the case of a reed-stop it is unnecessary to perforate the pipe, as

[1] Bonavia-Hunt, ' Modern Studies in Organ Tone ', *Musical Opinion*, 1933.

the reed forces the pipe to vibrate in the required partial. The trumpet and the tuba are frequently harmonic, at least in the upper register. Occasionally reed-pipes are made with the pipe too short for its vibrations to be in unison with the reed. This tends to suppress the fundamental and give a thin quality of tone.

Many organs contain a stop variously known as 'acoustic bass', 'resultant bass', &c. It uses the phenomenon described on p. 28, obtaining a note of low pitch as a difference tone between two notes of higher pitch. Its practical advantage is that very large pipes can be dispensed with if we are prepared to accept a somewhat inferior power and quality in the required note. The loss is not so serious as one might expect. If two notes making an interval of a fifth are strongly sounded together, then a difference tone an octave below the lower note is heard. Thus a note of 16-foot pitch combined with one of 10⅔-foot pitch gives a note of 32-foot pitch without requiring a 32-foot pipe. This is best accomplished by a compound stop of two ranks. The device was introduced in a practical form by Abbot Vogler (1749–1814).

Analysis of Organ Tone.—On the analysis of organ tone a great deal remains to be done. The result of the analysis is apt to depend on the position of the microphone used to pick up the sound and on several other factors. It is important that all these variables should be eliminated, in order that we may be sure that the analysis is characteristic of a stop, and not merely of an individual pipe. This is all the more necessary since there is no objective physical standard of diapason tone, flute tone, or string tone. A beginning has been made by Boner.[1] An analysis was made of the tone of each of four pipes by different makers from each of six stops. The four pipes from a single stop showed individual differences due to differences in design. For instance, in the case of the viole d'orchestre the specifications were as follows:

Pipe.	Dia-meter, in.	Mouth width, in.	Cut-up, in.	Tuning.	Nicks per in.	Bearded.	Pres-sure, in.	Weight.
A	0·88	0·60	0·22	Slit	50	Yes	4	0·40
B	0·75	0·53	0·22	Slit	50	Yes	4	0·45
C	0·90	0·75	0·24	Slit	27	Yes	5	0·65
D	0·94	0·63	0·22	Slit	25	Yes	6	0·58

[1] *Journ. Acous. Soc. Amer.*, Vol. 10, p. 32 (1938). Lottermoser has carried out a considerable amount of work in Germany, and reports may be found in the *Akustiche Zeitschrift* and the *Zeitschrift für Naturförschung*.

Nevertheless the individual differences in the results of the analysis did not altogether obscure a general resemblance, which justified taking an average for the four pipes for each stop and regarding the result as at least generally characteristic of the

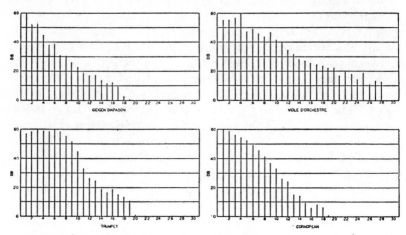

FIG. 8.13.—Analysis of tone of four organ stops. The average for four pipes of each stop was taken

stop in question. Fig. 8.13 shows the results for the four stops. The length of the ordinate indicates the strength of the corresponding partial. The results may be summarized as follows:

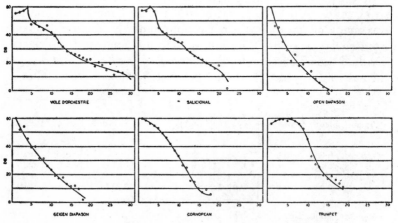

FIG. 8.14.—Analysis of tone of six organ stops shown by drawing smooth curves through the ends of the ordinates in each case

(*a*) Geigen Diapason—first partial predominant; second and third fairly strong; partials up to the nineteenth perceptible.

(*b*) Viole d'Orchestre—first four partials strong; fifth to ninth fairly strong; partials up to the twenty-eighth show some strength.

(*c*) Trumpet—first seven partials all very strong, then a fairly rapid drop in strength to the twentieth.

(*d*) Cornopean—gradual diminution in strength from first partial to sixteenth partial.

If an attempt is made to smooth out minor irregularities and draw a smooth curve through the ends of the ordinates for the six stops tried, the results are shown in Fig. 8.14. The similarity between the geigen diapason and the open diapason is interesting, as showing a similarity in quality.

CHAPTER 9

SUNDRY OTHER MUSICAL INSTRUMENTS

The Human Voice.—The sounds of the human voice are produced by a stream of air forced through a double reed which is coupled to a series of air cavities. The double reed—known as the vocal chords—is situated just above the junction of the wind-pipe with the larynx (Fig. 9.1). It is in effect a pair of lips, not unlike the lips of the human mouth. The edges can be brought together and air from the lungs forced through between them. They will then be set in vibration, and will produce a periodic interruption of the air-blast, behaving much as do the lips of a bugler or cornet-player. This gives the fundamental of the note that is sung or of the speaking voice, and the production of sound by using the vocal chords is known as phonation. The pitch of the note produced depends on the thickness, tension, and vibrating length of the two 'chords', factors which can be varied. The two chords lie in the same horizontal plane, and run from front to back. They open widely during the act of

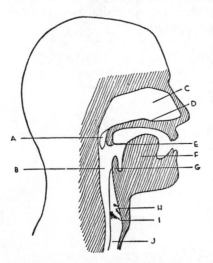

FIG. 9.1.—Diagram of vocal chords and associated cavities. A, soft palate; B, pharynx; C, nasal cavity; D, hard palate; E, uvula; F, tongue; G, epiglottis; H, false vocal chords; I, vocal chords; J, wind-pipe

breathing, and are about 2 cm long in the case of men and about $1\frac{1}{4}$ cm long in the case of women. The 'cracking' of the adolescent boy's voice is due to a comparatively rapid change in the length of the chords to about twice the previous value.

The air from the vocal chords passes up through the pharynx across the upper surface of the tongue and out through the lips, or

PLATE IX

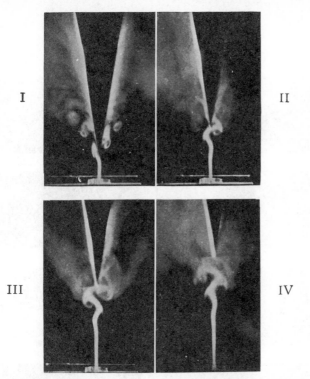

FIG. 8.6.—Four stages in vortex formation producing the same frequency

FIG. 8.4.—Eddies formed by drawing an obstacle through still water

PLATE X

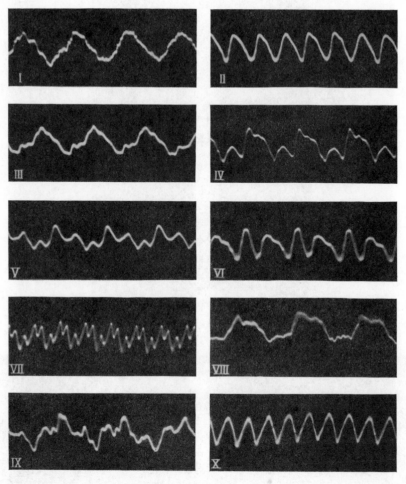

FIG. 8.11.—Oscillograms for vibrations of air in a pipe blown at different pressures

    I. Underblown fundamental.
    II. Underblown octave.
    III–VII. Fundamental with partials.
    VIII–IX. Fundamental and partials for closed pipe.
    X. Overblown twelfth for closed pipe.

behind the soft palate into the nasal cavity, and out through the nostrils. The manipulation of the tongue and lips produces notable changes in the natural frequencies of the air cavities, and so modifies the quality of the note, the various vowel sounds being produced in this way.

The male speaking voice has an average fundamental frequency of about 145 Hz, with a range of about 12 tones; while the female speaking voice has an average frequency of about 230 Hz, with a similar range. The lowest note of a cathedral choir bass has a frequency of about 66, while the highest note of a soprano is about 1,056, or about four octaves higher.

The loudness of the voice depends mainly on the pressure at which air is forced between the vocal chords. It varies with pitch, both for men and women, and the power is greatest near the top of the range, where it may rise to about 1 watt, which may be compared with the figures given on p. 34.

Attempts have been made by W. T. Bartholomew to ascertain the physical characteristics of ' good voice-quality ' in male singing voices.[1,2] He finds that there are four such characteristics: (1) satisfactory vibrato, (2) satisfactory loudness without strain, (3) marked low formant, (4) marked high formant.

(1) Vibrato is responsible for ' life ' or ' warmth ', and has already been discussed (p. 56). Good voice-quality seems to be inseparably connected with a smooth and fairly even variation of frequency, 6 or 7 Hz, and involving pitch, loudness, and quality. The vibrato is not necessarily heard as such. It is appreciated as a certain richness of tone and accepted as part of the timbre or quality.

(2) Loudness produced without effort seems to be associated with a relatively enlarged throat, allowing vigorous action of the cords and giving a free passage to the resulting sound. Examination of the voice-organs in the case of poor voices usually shows a constricted pharynx and partially closed epiglottis.

(3) Good voices emphasize frequencies round about 500 Hz, and this gives the voice its ' resonance ' or ' roundness ' or ' sonority '. This formant seems to be the natural frequency of the pharynx. It may be secured by emphasizing the lower of the two vowel formants, and there is some evidence that singers sacrifice the enunciation of their vowels to the demand for good tone-

[1] *Journ. Acous. Soc. Amer.*, Vol. 6, p. 25 (1934).
[2] See also Seymour, *Acustica*, Vol. 27, p. 203 (1972).

quality by modifying the formant required for certain vowel sounds so as to bring it down to about 500.

(4) The high formant in male voices lies between 2,400 and 3,200. It contributes the 'ring' or 'shimmer'. It is almost universally present, although it is much stronger in some cases than in others. It is heard even when the pitch is as low as 110, although in this case it is the 25th partial. Its universal presence seems to suggest that it is due to some relatively rigid part of the voice-organs, and a part of the larynx is suggested in the paper quoted. This suggestion is supported by the record in Fig. 9.2, Plate XI. This record was obtained by blowing the excised larynx of a calf. The lower curve gives the time-scale, the upper curve shows that each puff of air that comes from the vocal chords induces a high-frequency vibration, which is almost completely damped out before the next puff of air is released by the chords. It is an interesting fact that the frequency of this high formant in the male voice is close to the natural frequency of the meatus and to the frequency to which the ear is most sensitive.

The whole investigation is illustrated by Fig. 9.3, Plate XI, in which a and b are records of a voice of good quality and c and d of a voice of poor quality. There is a time interval of about $\frac{1}{13}$th of a second between each pair so as to reveal the vibrato, if any. The marked difference between a and b shows it to be present; the similarity between c and d shows it to be absent. The greater amplitude of b shows it to belong to a tone of greater intensity. The narrow kinks in a and b reveal the existence of a strong high-frequency component.

Fig. 9.4 shows an analysis of a record for the singing of the Bach–Gounod 'Ave Maria, gratia plena, Dominus tecum', given by Seashore,[1] who analyses it as follows:

'The song is in the key of F. The first note on the syllable *A* was approached from below, rose above the mean pitch, and then remained on true pitch for a little over 3 seconds. The *ve* began in true pitch, but rose to about a quarter tone sharp for three-fourths of the note. The *Ma* was touched but lightly on two vibrato cycles, followed by an upward glide on the third cycle. The *ri* was begun a trifle flat, but after three cycles rose to true pitch, and remained true up to the portamento to the second note on the tie; it then flirted

[1] *Psychology of Music*, p. 269.

with this note on two wide cycles and glided upward until it was continued on the syllable *a* in true pitch throughout this note.   After a breath-pause the note on *gra* was attacked by a two-tone rising glide, followed by two cycles, slightly

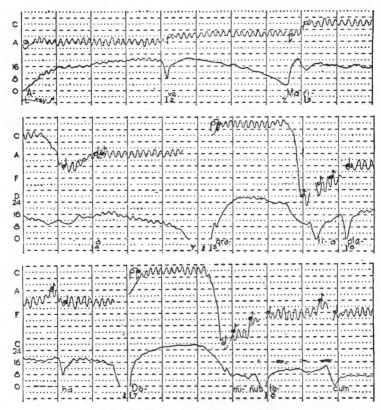

From Seashore's *Psychology of Music* (McGraw-Hill).

FIG. 9.4.—Record of loudness and pitch (Bach-Gounod, 'Ave Maria').   The upper curve shows variations of pitch; the lower curve shows the corresponding variations of loudness

flat, reaching the true pitch about the third cycle, from which there was a gradual flatting followed by a gradual return to true pitch, the portamento to the second note of the interval being considerably overreached.   The movement from the flatted second note on the tie continued on the next two notes, *ti* and *a*, under a relatively even upward glide in which the true notes were only touched by the crest of the very wide

vibrato cycles. *Ple* was sung on approximately true pitch
for the first note of the tie, and only the crest of the vibrato
cycle reached the second note of the tie from which a downward glide carried it to the true pitch of the note on *na*, which
was then sung on approximately true pitch.

'After a breath-pause, the note on *Do* was attacked two
and one-half tones from below and rose leisurely for 0·7
second to true pitch, which was held to the glide on the second
note, which was slightly overreached, so that on this and the
next two notes, *mi* and *nus*, we have a parallel to the three
notes at the end of the fifth measure. Both notes for the tie
on *te* were sung flat but a half tone short. The note on *cum*
was attacked flat but was held at true pitch after the first
two cycles. The first note in the ninth measure reached
the true pitch in three cycles, began the slide gradually,
and slightly overreached the second note. From this note
through the next five notes we see the characteristic figure of
the legato movement, that is, the notes lose individuality and
are blended in a gliding, relatively smooth inflexion which
constitutes a natural unit.'

It will be noticed that the pitch of a note varies a good deal
about the true pitch as mean. Frequently the pitch departs from
the true value considerably, the errors being greater for short
notes than for long ones. Long notes tend to begin slightly flat
and rise to true pitch. The notes in the upper and lower registers
tend to be flat, those in the middle register tend to be sharp. Of
the 107 notes, 48 were sung level, 37 rising, 9 falling, and 6 erratic.

The Flute.—The ordinary transverse flute as we know it is
similar in principle to the flue-pipe of the organ, and consists of
an edge tone coupled to a pipe tone. The lips play the part of
the flue of the organ-pipe, the air being forced through them in a
flat jet. The edge of the embouchure or blowing-hole gives the
edge tone, and the performer can vary the distance of his lips from
the edge and the velocity of the air-stream, and so control the
edge tone. The embouchure is cut near one end of the pipe, which
is cylindrical and closed at this end with a stopper. The end
remote from the embouchure is open. Originally the pipe was
in one piece, and was pierced with six finger-holes of suitable size
and appropriate position. When these were all closed and the
edge tone tuned to the fundamental of the pipe, the frequency was

that associated with a wave-length twice the effective length of the tube. The effective length is the actual length from the centre of the embouchure to the open end, together with the end corrections for both ends made necessary by the fact that the anti-nodes are not exactly at the embouchure and open end respectively but some distance beyond (see p. 126). Opening the end-hole shortened the effective length of the pipe and gave the next note of the scale. The simple theory is shown in Fig. 9.5. To get the

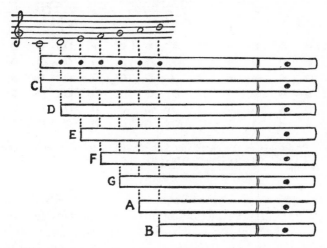

FIG. 9.5.—Diagram showing the production of the scale on the flute in C. The effective lengths of the vibrating air columns are shown at C, D, E, &c., as the corresponding holes are opened

next octave all that is necessary is to adjust the edge tone to the second partial of each length. In the same way the third and fourth partials may be elicited to give the next octave. In this way a complete scale in the key of the fundamental for the whole length of the flute and extending over three octaves becomes possible. It makes no provision for the chromatic scale, however. This difficulty is met by the fact that the position of the antinode at the farther end from the mouth depends not only on the position of the open hole nearest to the mouth (which approxi-mately defines the position of the antinode), but also on whether the holes beyond it are open or closed. A closed hole beyond it gives less freedom for the vibrating air and displaces the antinode farther from the embouchure, increasing the wave-length and lowering the pitch. Using this method of fork-fingering, the

chromatic scale is achieved as shown in Fig. 9.6. This provides
the complete chromatic scale except for *d'♯*. The quality of the
notes produced by fork-fingering was generally regarded as unsatisfactory, but no important improvement was made so long as
the flute remained keyless. A surviving specimen is divided into
three parts; the head joint contains the mouth-hole, a middle
piece has no hole, and the end piece has six finger-holes.

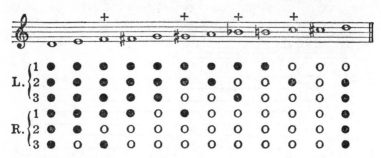

FIG. 9.6.—Production of the chromatic scale on the flute in D by fork-fingering;
black dots indicate closed holes, circles indicate open holes. The forked sounds
marked thus +

Further advance was made possible by the introduction of
keys. These function as additional fingers, closing holes in
addition to the number which the fingers can close; as longer
fingers, reaching holes otherwise inaccessible; and as thicker
fingers, closing larger holes than the fingers can effectively close.
The simplest form of mechanism is shown in Fig. 9.7. In the case

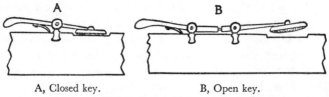

A, Closed key.   B, Open key.
FIG. 9.7.—Simple forms of key used in the flute

of the closed key the hole remains closed until the finger is applied
to the key; in the case of the open key the opposite is true. The
mechanism of these keys has been steadily improved, and methods
for the simultaneous manipulation of two or more with one finger
have been devised.

A key was introduced early in the seventeenth century closing a hole bored below the E hole.   This gave the missing $d'\sharp$.   In this form the instrument was used during most of the eighteenth century, and survived into the nineteenth century.   Its bore was slightly conical, it was divided into three or sometimes four pieces, and its fundamental was d'.   Soon, however, other keys began to be introduced to give semitones without fork-fingering, until the four-keyed flute of the Beethoven period.   Later still two open keys were introduced as a further extension, so that when the keys were closed, two semitones lower than d' could be obtained. Further additions gave the eight-keyed flute in use about 1820. At this date Theobald Boehm appeared on the scene.   He had been trained as a jeweller, he was a proficient flute-player, he had a distinct scientific bent and some inventive genius.   He set to work to improve the flute by disregarding completely the limitations imposed by the human hand owing to the limited number, size, and range of the fingers.   Holes were put where the tone of the flute required them, and were made of the size found to give the best results, the fingers being assisted by keys wherever necessary.   Starting with the tapering flute, Boehm substituted a cylindrical pipe with a slightly tapering head.   The sounding length was 57 cm., with a diameter of $\frac{1}{30}$ of this—i.e., 1·9 cm.— tapering to 1·7 cm. at the stopper.   The work of Boehm has been thus described by Redfield.[1]

When Theobald Boehm died at the age of eighty-eight there was little left to identify the instrument with which he began.   The instrument, which theretofore had been con- structed of wood, was changed into one of silver.   The bore of the instrument, which had been conical, increasing in diameter from the lower end to the upper, he changed so that it became cylindrical.   The inside dimensions of the head-joint he completely revolutionized and he changed the distance of the blow-hole from the end plug.   He changed the height and design of the posts which support the keys and improved the screws entering the posts as well as the method of fastening the springs to the body.   He so revolu- tionized the system of fingering used in playing the instrument that flautists had to learn to play all over again.   He changed the manner of blowing the flute, and the character of tone produced by it.   Indeed the only feature of the flute I can

[1] *Music : a Science and an Art*, A. Knopf.

think of that Theobald Boehm did not change was the position in which it was held to be played.'

Among other instruments of the same class are the piccolo and the fife, in both of which the air-stream is directed by the lips on to the edge of the embouchure, and the recorder and flageolet, both of which have whistle mouth-pieces. King Henry VIII is said to have possessed nearly 400 musical instruments, of which 154 belonged to the flute family. He himself played the recorder, of which the inventory of his estate lists seventy-six.

The Wood-winds.—The wood-wind instruments form a very important class in the orchestra. The quality of tone is characteristic. Its effect on Pepys has been described by him in his Diary.

> 'That which did please me beyond anything in the whole world was the wind-musique when the angel comes down (in the play 'The Virgin Martyr') which is so sweet it ravished me, and indeed, in a word, did wrap up my soul so that it made me really sick . . . that neither then, nor all the evening going home, and at home, I was able to think of anything, but remained all night transported, so as I could not believe that ever any musique hath that real command over the soul of a man as this did upon me; and makes me resolve to practise wind-musique and make my wife do the like.'

All the members of this class have a reed—usually made from a real reed—associated with a pipe. The clarinet, basset horn, and saxophone have a single reed, the oboe, English horn, and bassoon a double reed. The clarinet is nearly cylindrical, the oboe nearly conical. The single reed is flat, and very thin and flexible at the end which is placed in the player's mouth. It behaves like the striking-reed of the organ-pipe. In the case of the double reed, two flat pieces are bound firmly round a small metal tube at their thicker ends, and leave a small aperture between them at the free end where they are held between the player's lips.

The simple theory of the clarinet treats it as a coupled system, consisting of a cylindrical pipe closed at one end and a reed. If this were strictly true, the analysis of clarinet tone would show the absence of the even partials. Analyses have been made by various observers, and all agree that the second and fourth partials are relatively weak. They are by no means negligible. however, and in some analyses the eighth and tenth partials are especially prominent. This is probably due to the fact that neither of the assumptions ordinarily made is quite justified. On the one hand, the tube is not cylindrical. It has a pronounced flare at the foot, and the mouth-piece is very irregular in shape. On the other hand, the reed end is not strictly closed, the air-stream never

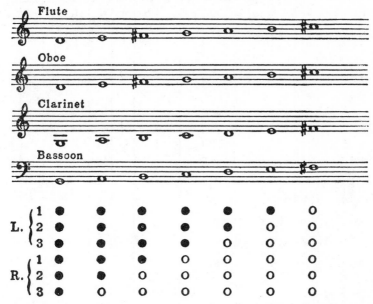

FIG. 9.8.—Production of the scale on the instruments shown; black dots represent holes closed, rings indicate holes uncovered by fingers

being completely stopped, and the vibrations in the tube communicating themselves through the reed opening to the air in the mouth and chest of the player.[1]

The scale is produced in the same way as in the case of the flute, as will be seen from Fig. 9.8, and there has been the same history of development by adding keys. In the case of the clarinet more keys are required to compensate for the fact that when over-blown it gives the twelfth, and not the octave.

The quality of tone of the clarinet may be associated with the predominance of odd partials and the relative prominence of the

[1] See, for example, Chatterji, *Proc. Nat. Acad. Sci. Ind.*, Vol. A21, 261 (1952); or Ghosh, *Journ. Acous. Soc. Amer.*, Vol. 9, p. 255 (1938). Some authorities are of the opinion that the reed does completely close (McGinnis and Gallacher, *Journ. Acous. Soc. Amer.*, Vol. 12, p. 529 (1941)) and the problem does not seem to be finally resolved

higher even partials.　In the case of the oboe the analyses suggest
a quite definite formant, fairly widely spread between the frequencies 600 and 1,500.　A typical set of analyses for the bassoon
is given in the table below taken from Seashore:

Bassoon—Percentage of Energy in each Partial

Partials.	1.	2.	3.	4.	5.	6.	7.	8.	9.	10.	11.	12.	13.	14.
C-523 f .	87	9	4											
p .	96	4												
G-392 f .	41	50	4	5										
p .	84	14	1	1										
E-329 f .	40	29	25	5										
p .	71	22	7	1										
C-262 f .	2	96	1	0	1									
p .	5	95												
G-194 f .	1	88	10	1										
p .	1	79	19	1										
E-163 f .	0	10	87	2	0	1								
p .	0	12	86	1	0	1								
C-130 f .	0	8	58	23	10	0	0	0	1					
p .	4	14	52	29	1									
G-97 f .	1	1	7	25	59	7								
p .	2	2	4	62	25	5								
E-82 f .	2	0	9	6	9	49	23	1	0	0	1	0	0	1
p .	11	3	2	16	4	42	2	1	16	1				

Here again there is evidence of a very definite formant.　Thus
if we take the frequencies of the prominent partials, we find the
following:

Note.	Frequency of prominent partial.	Note.	Frequency of prominent partial.
C-523 . . .	523	E-163 . . .	489
G-392 . . .	392	G-130 . . .	390
E-329 . . .	329	G-97 . . .	388–485
G-262 . . .	524	E-82 . . .	492
G-194 . . .	388		

Thus the prominent partials all have frequencies lying between
329 and 523.

Brass Instruments.—This group contains the French horn, the
trumpet, the cornet, the trombone, and a number of other important instruments.　They have certain features in common.

They all have a cup or funnel shaped mouth-piece to which the lips of the performer can be applied.   Air is driven through between the lips, the tension being controlled so that the lips act as a double reed coupled to the pipe.   The vibration of the lips may be seen in the photographs in Fig. 9.9, Plate XII.   The pipe is brass, either all more or less conical, or with the first portions cylindrical and the remainder more or less conical, and ends in a fairly pronounced flare or bell.   The length of the pipe can be varied—not by effectively shortening by opening holes, as in the case of the wood-winds—but by effectively lengthening it by inserting an additional piece of tubing, as in the case of the horn, or by the movement of a sliding piece, as in the case of the trombone.

Considering first of all the French horn, we find that it has a long tube, a conical and relatively narrow bore, a funnel-shaped mouth-piece, a large bell, and that the tube is coiled on itself. The tube is about 4 m long and the fundamental is F_1, though frequently, the instrument nowadays has Bb_1 as fundamental.   If there were no means of altering the effective length of the tube, then the notes available would be:

	F.	G.	A.	Bb.	C.	D.	Eb.	E.
F_1–F . .	1	—	—	—	—	—	—	—
F–f . .	2	—	—	—	3	—	—	—
f–f′ . .	4	—	5	—	6	—	7	—
f′–f″ . .	8	9	10	11	12	13	14	15

Thus in the fourth octave above the fundamental we have the complete diatonic scale with the addition of $e''b$, except that calculation shows 11 is a trifle too sharp and 13 a trifle too flat. If we can alter the effective length of the tube, then we can fill in the full chromatic scale in this octave by an alteration equivalent to a semitone shift, and by a larger shift we can fill in the full chromatic scale for the third octave.

We already know that the effective length of a tube can be increased by obstructing the vibrations of the air at its open end, and this method has been used with the horn.   The hand can be inserted into the bell, effecting a lowering of pitch of a semitone or even of a whole tone.   This, however, affects not only the pitch, but also the quality, and the method has now been superseded.

An alternative method is to have a removable ' crook ' which forms part of the tube, but which can be removed and replaced by a longer crook if desired.   This method has also been used, but has developed into a system in which three crooks are permanently attached to the instrument, any one of them or any combination of them being brought into operation by valves worked by keys.   This idea is shown in Fig. 9.10.   Ordinarily the sound-waves travel straight down the tube (coiled on itself in the actual instrument), but any one of three valves can be made to close the direct route and compel the sound-waves to travel round the valve-tube instead, and the valves can be operated singly or

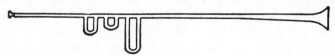

FIG. 9.10.—Diagram showing the principle of the valves used in a brass instrument. The length of the pipe can be effectively lengthened by throwing in at will any one or any combination of three additional lengths

in combination.   Using the second or middle valve gives a lowering of one semitone, the first valve gives a lowering of a tone, the first and second valves give a minor third, the second and third give a major third, the first and third give a perfect fourth, and the first, second, and third give a diminished fifth.   By the use of this system we get a complete chromatic scale from a diminished fifth below the second partial (or octave of the fundamental) as far upwards as it is possible to play.

It is conventional to assume that the instrument has

as its second harmonic; the scale is then given as follows:

FIG. 9.11.—Combinations of valves required to give the chromatic scale
The hollow notes are the open notes, Nos. 2, 3, 4, 5, 6, and 8 of the harmonic series.   No. 7 is not used because it is out of tune with the tempered scale

Three valves are used for the trumpets, cornets, and horns which do not make use of the fundamental; but for the euphoniums, tubas, and bombardons a fourth valve is added giving a lowering of a perfect fourth.

The use of valves in combination presents a difficulty which will become obvious on reflexion.   To lower the note of a pipe by a semitone requires an additional length which is a specified fraction of the length of the pipe.   If the valve-tube is the correct length for lowering the pitch of the pipe by a semitone, it will not be quite long enough to do so if the pipe is first effectively lengthened by the use of one of the other valves—i.e., the note will be too sharp.   In the shorter instruments this effect may be compensated by making the third valve-tube a little too long, as it will be in combination with the third valve that the effect will be greatest. In the longer instruments the difficulty can be met by arranging the valves so that extra compensating tube length is brought into operation when valves are used in combination.   Two types of valve are in common use—the piston and the rotary.

The quality of the notes of a brass instrument is governed by the same considerations as that for any other wind instrument and depends on

> (a) the shape of the tube,
> (b) the scale of the tube,
> (c) the bell.

In the case of the horn and most other brass instruments, the tube is approximately ' logarithmic ' or ' exponential ' in form.[1] This means that as we proceed by equal distances along the axis the radius of the section increases in a constant ratio, and therefore the area of the section also increases in a constant ratio.   Consequently the area doubles itself at equal intervals measured along the axis, and we can measure the flare either by (a) the ratio of the radii of cross-sections at intervals of one foot or (b) the distance along the axis in which the area doubles itself.   The properties of this type of tube or horn are further considered later (p. 210).   Meantime it may be noted that this particular form of horn gives a more efficient transmission of sound to the atmosphere, and that to give a good bass the flare must not be too rapid.   So far as the scale— i.e., the ratio of width to length—is concerned, the narrower the tube the more brilliant the quality.   Most instruments end in a bell in which the expansion is very rapid, and a large bell tends to give mellowness, while a smaller one tends to brilliance.

The analysis of the tone of the horn given by Seashore [2] is

[1] For instruments of the trumpet or trombone families, the best bell shape is found to be hyperbolic.
[2] *Psychology of Music*, p. 190.

shown in the following table. It shows a formant in the region between 200 and 600—i.e., *a* to *e″*, and Seashore's comment is ' the wide and well balanced spread of partials in this region gives the rich and mellow characteristic. The fundamental is practically absent below 150.'

The French Horn

Partials.	1.	2.	3.	4.	5.	6.	7.	8.	9.	10.
B-466 *f*	90	9	1							
p	86	12	2							
A-440 *f*	99	1								
p	26	73	1							
F-349 *f*	66	29	4	1						
p	94	6								
A-220 *f*	26	31	26	5	9	2				
p	77	6	14	2						
F-173 *f*	14	32	46	7	1					
p	10	43	36	9						
C-130 *f*	1	19	21	48	4	5	2			
p	9	30	25	30	5	1				
A-110 *f*	2	22	34	6	21	3	1			
p	11	34	4	25	11	9	4	1	1	
F-87 *f*	1	43	22	19	3	6	4	1		
p	0	12	7	10	15	15	27	8	3	2

In the case of the trumpet the physical principles involved are much the same. The old trumpet remained cylindrical from the

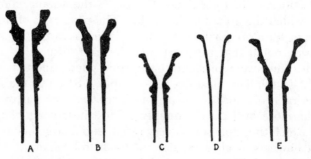

FIG. 9.12.—Different forms of mouth-piece for brass instruments

mouth-piece to a point about 45 cm from the bell-mouth; in the modern instrument the tube tapers slightly to the mouth-

piece.   There is also a difference in the shape of the mouth-piece
shown in Fig. 9.12.   This difference was more marked in the case
of the old trumpet, where the mouth-piece was hemispherical and
joined the tube in a sharp edge, in contrast to the conical form
of the French horn.   Richardson [1] suggests that the edge of the
mouth-piece where it joins the tube generates an edge tone which
contributes to the distinctive quality of tone.   Just as in the case
of the flue-pipe of the organ a straight edge presented to a flat jet
of air generates an edge tone, so it is known that a jet of circular
section meeting a circular edge generates a similar tone.   On

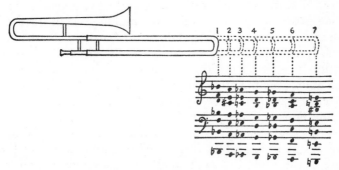

FIG. 9.13.—Diagram showing the partial tones given by the trombone for the
various positions of the slide

this view the trumpet is a system in which the reed tone of the
lips, the edge tone of the jet, and the tone of the pipe are coupled
together and vibrate as a coupled system.

In the trombone the variation in length of the tube is produced
by a slide.   The notes corresponding to the various positions are
indicated in Fig. 9.13.   This method of adjusting the length
enables the performer to produce correct intonation, but is, of
course, slower in action.

Vibrations of Bars.—If a pin be stuck into a soft piece of wood
and plucked, it vibrates transversely—i.e., at right angles to its
length—and gives a note.   A row of pins stuck in to varying
depths may be tuned so as to give the notes of the scale, as
the frequency of the transverse vibration depends on the length.
This primitive instrument may be regarded as the ancestor of the
musical box, in which a series of small bars form the teeth of a
metal comb and are set in vibration by studs projecting from a

[1] *Acoustics of Orchestral Instruments*, Edward Arnold, 1929.

revolving drum.   It is also possible to elicit longitudinal vibrations
—i.e., vibrations executed in the direction of the axis of the bar—
but these have no musical importance.   Almost the only musical
instrument depending on the vibration of bars is the xylophone.
In this case the bars are not clamped or fixed at one end, as in the
musical box, but are a graduated series resting on supports, and
prevented from slipping by screws fixed to the supports and passing through perforations in the bars.   The bars are made of
hard wood, and are hit with hammers.   The supports are about
one-sixth of the length from either end, and the fundamental
mode of vibration of the bar has nodes at the two points of support,
with antinodes at the ends and in the middle.   Thus the greatest
bending takes place at the middle of the bar, and the stiffness of
the bar is most affected by thinning it in this region.   If we lessen
the stiffness we tend to slow the vibrations and flatten the pitch
of the note.   On the other hand, if we thin the bar at its ends we
reduce the mass which the stiffness has to move, and so speed up
the vibrations and sharpen the pitch.   Thinning the bar at the
centre does, of course, reduce the mass, but the effect on the
stiffness is more important there, while thinning at the ends does
reduce the stiffness, but as the end is in maximum motion, being
an antinode, the effect on the mass is more important than the
effect on the stiffness.

The xylophone seems first to have appeared in Europe as the
strohfiedel—a set of bars resting on straw bands and struck with
small hammers.   An illustration (Agricola, 1528) shows an
instrument with a compass of three diatonic octaves.   The xylophone is used with effect by Saint-Saëns in the ' Danse Macabre ',
although it is largely a virtuoso instrument.   Its partials are
few and high, the first five modes having frequency ratios
$1 : 2 \cdot 76 : 5 \cdot 4 : 8 \cdot 9 : 13 \cdot 3$, and its tone is therefore very pure—all the
more so as the bars are struck at points where the higher partials
require nodes.   Sometimes the strength and purity of tone are
increased by the use of resonators, as in the celesta.

The tuning-fork also depends for its note on the transverse
vibrations of a bar, and may be regarded as a bent bar in which
the two nodes approach one another as the bending is increased.
The stouter the shaft and base of the fork, the closer do the two
nodes approach one another.   The movement of the shaft of the
fork transmits the vibrations to any surface on which the shaft is
rested, or to the resonance box on which the fork is mounted.

The first three modes of vibration are shown in Fig. 9.14.   These are widely separated in pitch, and in one case the frequency ratio was found to be $1 : 6 \cdot 33 : 17 \cdot 07$.   The second and third partials form the ringing note, which is heard when the fork is struck on a hard surface.   If the fundamental is elicited by bowing or by striking with a soft hammer, the tone is nearly pure.

The tuning-fork owes its importance to its convenience and constancy as a standard of pitch. It was invented by John Shore, a trumpeter for Handel, and one of a group of twenty-four musicians who formed an orchestra for Queen Anne.   In 1714 he was Sergeant Trumpeter at the Coronation of George I,

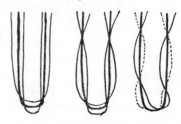

FIG. 9.14.—Modes of vibration of tuning-fork

and later he was Lutenist in the Chapel Royal.   He always carried a fork to tune his lute, and his stock joke is said to have been, ' I do not have about me a pitch-pipe, but I have what will do as well to tune by, a pitch-fork ! '

If we carry the bending process a step farther, we get the triangle, which gives numerous strong partials and no definite pitch.   It appears in 1775 in Grétry's ' La Fausse Magie ' and in 1795 in Haydn's ' Military Symphony '.

Vibrations of Stretched Membranes.—The vibrations of stretched membranes are used in the drums and the tambourine.   The drum has in general two membranes, mounted over the ends of a cylinder of wood.   One of the membranes is struck, and the other is set in vibration owing to the fact that the membranes are coupled by the enclosed air.   A study of the vibration of the two membranes shows that energy surges from one to the other, the membrane which is struck losing its energy to the other and almost ceasing to vibrate, and then gaining energy from the other, while it in turn ceases to vibrate.   This exchange goes on until the energy of the blow is exhausted.   In the military drum the lower membrane has a cord or wire stretched across it, called a snare. The membrane vibrates against the snare, and this is the cause of the peculiar timbre of these drums.

Most drums are mere rhythm-markers, but the kettle-drum is an   exception.   The membrane is stretched over a shell-shaped case, and is tuned by varying the tension.   It was introduced by

Lully about 1670, and three such drums were used by **Weber in** his Overture to ' Peter Schmoll ', 1807. Berlioz used eight pairs in his ' Requiem '. The air cavity ought to be of such size and shape that its natural frequency lies near to that of the membrane.

The possible modes of vibration are easily derived theoretically and demonstrated practically. In the fundamental the membrane vibrates as a whole, with an antinode at the centre and a nodal line round the rim. The frequency is inversely proportional to the diameter of the membrane, inversely proportional to the square root of the mass of the membrane per unit area, and directly proportional to the square root of the tension. The vibrations have been studied by Obata and Tesima, and the frequencies measured for a particular membrane, with the following result: [1]

Nodal pattern.	Frequency.	Ratio of frequencies of partial and fundamental.
None	214	1·00
One diameter	333	1·61
Two diameters	468	2·19
One circle	492	2·30
Three diameters . . .	544	2·54
One circle and one diameter .	621	2·91
Four diameters	695	3·25

The membrane is struck at a point some half-way between the centre and the circumference, and this eliminates the fourth partial, which has a nodal line at that point, and weakens several of the others.

Cymbals and Bells.[2]—Reference has already been made to the vibrations of plates and to the way in which the vibrations were studied by Chladni. These vibrations have no musical application in the case of simple plates, although the cymbals may be regarded as circular plates sunk at the centre. Circular plates differ from circular membranes in requiring no tension, but they form nodal patterns of the same kind, the nodal lines consisting of diameters and concentric circles. Since a handle is fixed at the centre of the cymbals, that point is a node.

The bell may be regarded as a further development of the plate. Bells are made in all sizes, varying from the small hand-bell to church bells weighing many tons. The largest bell in Europe is the Great Bell of the Kremlin. Its weight is estimated at 180

[1] *Journ. Acous. Soc. Amer.*, Vol. 6, p. 267 (1935).
[2] See also Grützmacher, *Acustica*, Vol. 4, p. 226 (1954).

tons, and its height is about 6 m. It was cast in 1734, and the clapper was pulled by twenty-five men on each side. In 1737 it fell, a fragment 2 m high and 11 tons in weight being broken off. It was partially buried by the fall, but was raised in 1836 and placed on a pedestal as a curiosity. The largest ringable bells are Great Paul ($16\frac{3}{4}$ tons) at St. Paul's Cathedral, and Big Ben

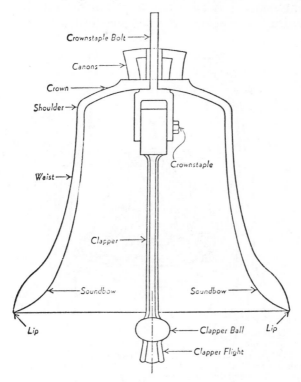

Crownstaple Bolt →

Canons →

Crown →

Shoulder →

Crownstaple

Waist →

Clapper →

Soundbow

Soundbow

Lip

Lip

Clapper Ball

Clapper Flight

FIG. 9.15.—Shape of modern bell

($13\frac{1}{2}$ tons) at Westminster. Bells are usually cast in an alloy of about 13 parts of copper to 4 parts of tin, and when intended for musical use a number are arranged to give some of the notes of a scale, and are known as a ring.

The best shape has been arrived at after centuries of experience, and bell-founding is a highly skilled art, practised by only a few firms. The form of the bell is shown in Fig. 9.15. The marked thickening of the bell at the sound-bow, where the clapper strikes, eliminates some of the high partials, and prevents the tinkling sound which would otherwise be very marked. When the bell is mounted for ringing it is attached to a wheel, as shown in Fig. 9.16. When the bell is about to be rung, it is pulled into an inverted position, and is then 'set'. A bell may be chimed by swinging it through a small arc, but ringing involves pulling it from an inverted position back to an inverted position again, and change-ringing is an art little practised except in England and America. In ringing a change each bell is rung once, the

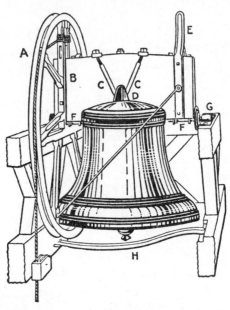

FIG. 9.16.—Method of hanging a bell for ringing. A, wheel with rope attached; B, headstock; CC, straps or keys; D, cannons (modern form); E, stay; FF, gudgeons; G, brasses (in which the brasses revolve); H, slider

order being varied according to certain rules. The record peal was made on April 9, 1960, at the church of St. Michael at Kirby-le-Soken by the Essex Association of Change Ringers, when 22,400 changes of Plain Bob Major were rung in 11 hours and 45 minutes.

An analysis of the sound from a church bell reveals some very interesting points. There are, of course, numerous possible modes of vibration, each with its appropriate partial tone. The simplest mode has two nodal meridians at right angles to one another. As we go round the sound-bow we find four nodes and four antinodes equally distributed, and this is true of any parallel circle above the sound-bow. The position at which the clapper strikes the sound-bow will determine the position of an antinode, and therefore of all the nodes and antinodes. It may happen, however, that the bell is not symmetrical, and in this case it will not be a matter of indifference where the clapper strikes the sound-bow. This possibility may be illustrated by using a tea-cup, in which case the symmetry is destroyed by the load supplied by the handle. If the cup is struck at a point opposite to the handle or at $90°$ either side from the handle, then the handle will be at an antinode and in maximum motion. If, on the other hand, it is struck $45°$ from the handle either side, then the handle will be at a node and will remain practically at rest. The note in this case is noticeably higher in pitch, the vibrations being more rapid, since they do not involve motion of the handle, which acts as a load without increasing the elastic forces which produce the motion. If the cup is struck at any intermediate point, both notes may be heard, and the corresponding phenomenon in the case of a bell, where the two frequencies may be very close, is sometimes the cause of beats.

The first systematic study of bells seems to have been made by Rayleigh in 1879 on the ring of bells at Terling. He found that the notes given by one of the bells were approximately

$$d'- \qquad a'- \qquad d''+ \qquad g''+ \qquad b''+$$

Thus the frequency ratios were approximately

$$1 \qquad 1\cdot 5 \qquad 2\cdot 0 \qquad 2\cdot 7 \qquad 3\cdot 3$$

He found that the ' strike '-note—the note which gives the pitch of the bell when it is heard in a series—was b', an octave below the fifth partial. He confirmed this on seven other bells. In every case the strike-note was almost exactly an octave below the fifth partial. On two bells it was near to the second partial, but on the others no partial was near it in pitch. Among others who have worked on this problem is A. T. Jones.[1] For the largest bell of the Carlisle Chime at Smith College, U.S.A., he found the frequency ratio of the first seven partials to be

$$1\cdot 00 \qquad 1\cdot 65 \qquad 2\cdot 10 \qquad 3\cdot 00 \qquad 3\cdot 54 \qquad 4\cdot 97 \qquad 5\cdot 33.$$

He found that the strike-note could not be picked up by a

[1] *Phys. Rev.*, Vol. 16, p. 247 (1920).

resonator, could not be elicited from the bell by resonance, and gave no beats with a note of nearly the same pitch. It does not appear to be a difference tone, nor does it appear to be due to a longitudinal vibration. He suggests that it is determined by the fifth partial, and that the octave is misjudged owing to the different rates of decay of the various partials.

In the case of chimes, the bells are hung 'dead', and struck by hammers. The hammers may be manipulated from a keyboard, but are usually set in motion by a revolving barrel with pegs or studs. The most famous chimes are probably the 'Cambridge Quarters' arranged in 1793 by Dr. Crotch for the University Church. They were copied at the Royal Exchange, London, in 1845, and at the Houses of Parliament in 1859. This last association has led to the name 'Westminster Quarters'.

A carillon consists of a large number of bells played from a keyboard with electrical action, which eliminates the need for heavy touch. They give considerable scope for melody. The home of the carillon is Belgium and Holland, and among the most famous are those of Malines (forty-five bells), Bruges (forty-seven bells), and Ghent (fifty-two bells). They have spread to England and America, one at Cattistock in Dorset having thirty-five bells, while the Riverside Carillon in New York has sixty-four bells, the largest of which weighs $18\frac{1}{4}$ tons.

As a matter of fact the material of the bell not only vibrates transversely (i.e., at right angles to the surface of the bell), but also longitudinally (i.e., to and fro round the surface of the bell). The antinodes of the transverse vibration are the nodes of the longitudinal vibration. This can be illustrated by a method due to Galileo. If a wetted finger is drawn round the ring of a finger-bowl or wine-glass, the note of the glass is strongly elicited. As the motion of the finger is round the rim, it is the longitudinal vibration which is elicited, and there is, of course, an antinode at the point at which the finger is being applied. If there is some liquid in the glass, a pattern will be observed on its surface from which it will not be difficult to see that the liquid, set in motion by the transverse vibration, shows an antinode at 45° from the point where the finger is being applied.

The use of a series of glasses as a musical instrument was introduced by an Irishman called Pockrick in 1743 and was known as the Angelica. Gluck performed a concerto for twenty-six glasses with orchestra in 1746. Benjamin Franklin made some improvements in 1760, and constructed the Harmonica or Armonica with thirty-three glasses. This instrument was for a time taken seriously. Mozart wrote an Adagio e Rondo for it with flute, oboe, violin, and violoncello, and Beethoven and others also composed for it.

The Pipeless Organ.—There has in recent years been a very considerable development of organs which are ' electrical ' in the sense that the notes are produced by electrical oscillations and transmitted to the hearers from one or more loud-speakers. The technical details of these instruments vary enormously, as also do the principles on which they work, and any satisfactory account of them would occupy far more space than is permissible in a book of this size. The design of the instruments themselves is in the process of development, and a law of survival of the fittest is gradually reducing the number of those likely to be of permanent value. The fundamental principle common to all is that a variable electric current of any frequency and harmonic content can be produced by any one of a variety of methods. This variable current can then be amplified without distortion and fed to a loud-speaker which will radiate sound of the prescribed frequency and harmonic content and of any required loudness. Theoretically, this enables us to reproduce any type of instrument—not only the various stops of the organ—but exact reproductions of the quality of tone of other instruments as well. And there are other advantages which this method of producing sound would seem to possess. The control of loudness is more effective, and gives greater range than is possible in the traditional organ with its boxes fitted with louvres to open and shut—a device which affects quality as well as loudness. Bulk is likely to be less, as a small electrical circuit can take the place of a 32-foot pipe. Constancy of pitch can be achieved and all subsequent tuning made unnecessary. Finally, even the initial cost should be relatively small.

When these theoretical advantages are examined more closely, however, some of them turn out to be largely illusory. We *can* produce any frequency of electric oscillation, but to amplify a note compounded of various frequencies without some distortion is a very costly business, and no ordinary loud-speaker will reproduce the whole audible range of frequencies. Then again, we *can* give the note any harmonic content we like, but if the instrument is to be a commercial proposition we cannot give each note an unlimited number of harmonics. The analyses given for organ pipes (p. 130) indicate that for certain qualities of tone a very long series of partial tones is required. Boner found that in a comparatively cheap organ a Salicional pipe produced the first thirteen partials with amplitudes greater than 1 per cent., and the partials 14, 15, 16, 17, 19, 20, and 23 with smaller, but still audible, amplitude. No such range of partials has been attempted on any pipeless organ. One of the best known commercial types, the Hammond, uses the partials 1 to 8, omitting the seventh, and adjusts their intensities to correspond to the quality of tone to be reproduced.

With recent developments in integrated circuits, very complex circuits can be provided in a very small space and some organs can now attempt to mimic the transient behaviour of organ pipes. Such instruments are very expensive and sometimes cost more than a traditional organ of equivalent specification.

CHAPTER 10

DISSONANCE AND CON-SONANCE

Dissonance.—When two notes are sounded simultaneously the combination of sounds is occasionally pleasing to the ear, but usually, if the notes are selected at random, the impression is one of unpleasant roughness, and may be described as dissonant. If we take two notes in unison and maintain the pitch of one of them constant while that of the other is gradually raised, we find that the dissonance goes through a series of maxima and minima. The dissonance has its first marked minimum when the note of variable pitch forms with the note of steady pitch the interval of the minor third. It passes through other minima at the major third, the fourth, the fifth, the minor sixth, the major sixth, and the octave. Some attempt to represent this graphically is shown in Fig. 10.1, which is based on a graph by Helmholtz. It will be noticed that the unison, the fifth, and the octave seem to be completely free from dissonance. It will also be noticed that the more perfect the consonance of an interval the more sharply it is bounded by dissonance, so that a mistuned minor third is much less dissonant

157

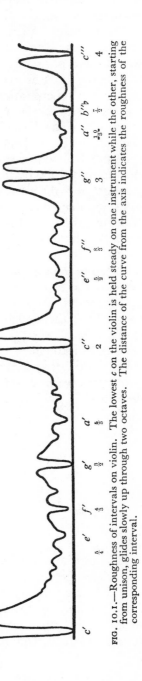

FIG. 10.1.—Roughness of intervals on violin. The lowest *c* on the violin is held steady on one instrument while the other, starting from unison, glides slowly up through two octaves. The distance of the curve from the axis indicates the roughness of the corresponding interval.

than a mistuned octave or a mistuned fifth—or, of course, a mistuned unison.

The physical explanation of dissonance was first suggested by Sauveur (1653–1716), and afterwards developed by Helmholtz. If we take two pure tones in unison and vary the pitch of one while that of the other is maintained steady, we hear 'beats' (p. 20) due to the superposition at the ear of the waves from the two sources. The beats are periodic variations of loudness, the frequency of the beats being the difference of frequency of the two tones. If the beat-frequency is anything up to five or six per second the effect is quite pleasing to the ear. As the difference in frequency of the two tones increases the beat-frequency increases and the sensation becomes less pleasant. Though the beats become too rapid to count there is no change in the *character* of the sensation, but the beats are now perceived only as a roughness. As the interval between the two pure tones is still further increased the roughness gets less, and at about a minor third the roughness disappears, and does not reappear until the interval between the two tones approaches the octave, when the roughness increases again, but disappears when the octave interval is exact. The theory advocated by Helmholtz is that dissonance in the case of *pure* tones is due to beats between the tones, the degree of dissonance depending on the beat-frequency and on the interval between the tones. Confining our attention still to pure tones, we find from the necessary calculations (p. 198) that the beat-frequencies for selected intervals are as follows:

Tones.					Interval.	Beat-frequency.
$c''-c''_{\#}$	.	.	.	.	Semitone	31·1
$c'-d'$	.	.	.	.	Tone	32·0
$c-e$.	.	.	.	.	Major Third	34·0
C–G	.	.	.	.	Fifth	32·5

Thus all these intervals have nearly the same beat-frequency, but while the semitone $c''-c''_{\#}$ is a very harsh dissonance, the fifth C–G is a consonance. On the other hand, comparing semitone intervals, we have the beat-frequency doubling every time we go up an octave, so that while for C–C$\#$ it is 4, for $c'''-c'''_{\#}$ it is 64. Of these semitones the most dissonant is $c''-c''_{\#}$ with 32 beats per second. Above this pitch the more frequent beating gives a smoother impression.

Experiments by Stumpf[1] give general confirmation of these phenomena and his results may be tabulated as follows:

Lower tone.		Maximum dissonance.		Dissonance disappears.	
Frequency.	Approx. pitch.	Beat-frequency.	Intervals in cents.	Beat-frequency.	Interval in cents.
96	G	16	266·9	41	615·7
256	c'	23	148·9	58	353·5
575	d''	43	124·9	107	295·5
1,707	a'''	84	83·2	210	200·9
2,800	f''''	106	64·3	265	156·6
4,000	b''''	—	—	400	165·0

Allowing for almost inevitable inaccuracies in measurements of this kind, we see that in the extreme bass the most offensive dissonance occurs for a beat-frequency of 16 per second and an interval between the two notes of just over a tone (one tempered tone = 200 cents). Dissonance ceases when the beat-frequency is 41 per second and the interval between the notes is a tritone (three tempered tones = 600 cents). In the middle range of pitch, d'', the worst dissonance has a beat-frequency of 43, corresponding to an interval of just over a semitone, while the dissonance disappears at about 107 beats per second, the interval then being about a minor third. For high frequencies, represented by f'''', the maximum dissonance occurs for 106 beats per second, the interval being just over half a semitone, and the dissonance disappears for an interval of about $\frac{4}{5}$ths of a tone, the beat-frequency then being 265.

The dissonance which occurs for pure tones an octave apart may be due to the formation of a difference tone. Thus if the frequency of the lower tone is f and that of the upper is $2f + \epsilon$, then the first-order difference tone is $f + \epsilon$, and this will make ϵ beats per second with the lower tone. It is probably these beats which cause the dissonance for an imperfect octave with pure tones. Our judgement must be tentative, for the beats are well marked only if the tones forming the interval are fairly intense, and in this case it is difficult to be sure that the tones are really pure. If not, the second partial of the lower tone would be $2f$,

[1] *Science and Music*, Sir James Jeans, Camb. Univ. Press, 1937, p. 50.

and this would make ϵ beats per second with the upper tone, so that the number of beats per second is the same in either case.

When we extend our study to the case of tones which are not pure, then the considerations which apply to a pair of pure tones apply to any pair of partial tones—so that any partial of the one tone can cause dissonance by beating with any partial of the other tone. In fact, in the case of a single note which carries a long series of partials, beating can occur between consecutive members of the same series, and roughness will be produced in this way. Between the eighth and ninth partials the interval is a tone only, and between subsequent pairs of partials the interval gradually diminishes.

When two musical notes are sounded together the resulting dissonance will therefore depend on:

(1) the interval between the pair of beating partials—generally speaking, partials will be in the higher frequency range, and beating will be worse for semitones than for tones;

(2) the strength of beating partials—this will depend on the quality of the notes, but in general the higher the order of the beating partials the less strong are these partials likely to be; it will also depend on the loudness of the two fundamental tones themselves.

Consonant Diads.—Defining a consonant diad as a pair of notes which, sounded simultaneously, produce a smooth, pleasant, and agreeable impression, we naturally think first of the *octave*. Next to exact unison the octave most nearly fits the definition. As we are no longer confining ourselves to pure tones, let us consider the notes as carrying the first six harmonic partials all sufficiently strongly developed to affect the ear. For the notes c and c' these may be set out as below:

We notice: (*a*) that the higher note merely reinforces what is already present in the lower note, and that so long as the tuning is accurate no possibility of unpleasant beating arises; and (*b*) that if the tuning is not accurate, beating occurs between every partial of the higher note and the corresponding partial of the lower note. This explains why the octave is so sharply bounded by dissonance.

The same considerations apply to the twelfth

Here there are in coincidence only two of the partials which matter, but otherwise the situation is unchanged. All intervals whose frequency ratio is a whole number come in this category— the octave (2 : 1), the twelfth (3 : 1), the fifteenth (4 : 1), &c.

It is interesting to note the effect of mistuning on the number of beats per second. For the octave, if the lower note is mistuned by one vibration per second, its second partial will be mistuned by two vibrations per second, and this will be the number of beats per second it will make with the higher note. Thus, if the correctly tuned octave is 512 and 256, and we mistune the lower note to give a frequency of 257, its second partial will be 514, and this will beat twice per second with 512. If, on the other hand, we mistune the upper note to 513, it will make one beat per second, with the second partial of 256, which is 512. We have here a particular example of a quite general rule. If we express the ratio of the frequencies of the two correctly tuned notes by the ratio of two small whole numbers, then the beats between the lowest pair of coincident partials for a mistuning of one vibration per second is the smaller number if the higher note is mistuned and the larger number if the lower note is mistuned. Thus for the fifth the ratio of the frequencies is 3 : 2, and the second partial of the higher note coincides with the third partial of the lower note, as shown below. It can easily be shown that these coincident partials give two beats per second if the upper note is mistuned by one vibration per second, and three beats per second if the lower note is mistuned by one vibration per second.

Looking at the fifth from the point of view of consonance, we can set out the partials as follows:

Here we notice distinct possibilities of dissonance: $c''-d''$ is a tone and $d''-e''$ is also a tone. The beating partials in the first case are the third and fourth, and in the second case the third and fifth. We have passed from the intervals where the upper note adds nothing which does not already exist in the lower, to an interval where the upper note adds new tones, and therefore new possibilities of dissonance.

Passing now to the fourth, we have:

$$c \quad\quad c' \quad\quad g' \quad\quad \genfrac{}{}{0pt}{}{c''}{c''} \quad\quad e'' \quad\quad g''$$
$$\quad\quad f \quad\quad f' \quad\quad\quad\quad\quad\quad\quad f'' \quad g'' \, a''$$

Here we have tone beating between f' and g', the order of these partials being 2 and 3 respectively, and again between f'' and g'' (order 6 and 4), and between g'' and a'' (order 6 and 5). But there is also semitone beating between e'' and f'' of order 5 and 4 respectively. Thus we should expect the fourth to be more dissonant than the fifth.

If we now take the remaining intervals in order from the note c in each case we have:

Interval							
Minor Third	c $e\flat$	c' $e'\flat$	g' $b'\flat$	$c.'$ $e''\flat$	e''	g'' g''	
Major Third	c e	c' e'	g' b'	c''	e'' e''	g''	$g''\sharp$
Minor Sixth	c $a\flat$	c'	g' $a'\flat$	c'' $e''\flat$	e''	g''	$a''\flat$
Major Sixth	c a	c'	g' a'	c''	e'' e''	g''	a''
Minor Tenth	c	c' $e'\flat$	g'	c'' $e''\flat$	e''	g''	$b''\flat$
Major Tenth	c	c' e'	g'	c''	e'' e''	g''	b''
Eleventh	c	c' f'	g'	c''	e'' f''	g''	
Twelfth	c	c'	g' g'	c''	e''	g'' g''	
Minor Thirteenth	c	c'	g' $a'\flat$	c''	e''	g''	$a''\flat$
Major Thirteenth	c	c'	g' a'	c''	e''	g''	a''

In the following table are shown for each diad the order of the partials which form an interval either of a whole tone or of a semitone, with consequent beating:

Interval.				Partials giving a tone.	Partials giving a semitone.
Minor Third	.	.	.	3–4	4–5
Major Third	.	.	.	—	3–4, 5–6
Fourth	.	.	.	2–3, 4–6, 5–6	4–5
Fifth .	.	.	.	3–4, 3–5	—
Minor Sixth	.	.	.	—	2–3, 3–5, 4–6
Major Sixth	.	.	.	2–3, 4–6	—
Octave	.	.	.	—	—
Minor Tenth	.	.	.	—	2–5
Major Tenth	.	.	.	—	—
Eleventh	.	.	.	1–3, 2–6	2–5
Twelfth	.	.	.	—	—
Minor Thirteenth	.	.		—	1–3, 2–6
Major Thirteenth	.	.		1–3, 2–6	—

We can set these intervals out in staff notation, using minims for the fundamentals and crotchets for the other partials, thus:

For the first six intervals we may now consider the order of dissonance on the assumptions that (a) semitone beating is more dissonant than tone beating, and (b) the lower the order of a beating partial the more dissonant is the result. Clearly the most consonant diad is the fifth. No interval has fewer pairs of beating partials. In the case of the major sixth the partials are of lower order, while in the case of the minor third one, and in the case of major third two of the beating intervals are semitones. On paper we should be inclined to put the major sixth and minor third next, followed by the major third and the fourth, with the minor sixth obviously last. In comparing this order with the order of aesthetic choice based on musical experience, we must remember the rather rough assumptions on which it is based, and remember also that we are treating these diads as studies in still life, while the musician assesses them as elements in a progression, some degree of dissonance having occasionally a positive function to fulfil, as when a discord is ' resolved '. There is universal agreement that the fifth is at the head of the series. Equally there will be agreement that the minor sixth comes at the other end of the series. There has always been uncertainty about the others, which may very well be due to the fact that their relative dissonance depends on the quality of the notes, and therefore on the instrument on which they are played. The fourth may be thought to come too low in the series, but it has always been a subject of dispute, and probably owes some of the precedence which it has in musical practice to the fact that it is the inversion of the fifth. Equally the minor sixth may be thought to come too low in the series, but its character depends, to some extent at least, on its use in the minor chord, which will be considered later.

There is, of course, no question at all about the theoretical dissonance of all other possible intervals—the minor and major seconds and sevenths and the augmented fourth or tritone.

Helmholtz attempted to calculate dissonance numerically for violin tone, and showed his results graphically, as in Fig. 10.1. It will be noticed how sharply defined are the consonances of the unison, the octave, the fifth, and the fourth. It will also be seen that for smoothness the major sixth compares favourably with the fourth.

Attempts have been made by von Békésy to establish a definite scale of roughness, but the work has not yet reached a satisfactory conclusion, despite much other work in the 1960's.

Consonant Triads or Concords.—Confining our attention to three notes within the compass of an octave, we require to find how these must be selected in order that they may produce an agreeable impression. This will obviously be achieved if we can add two consonant diads so that the extreme notes shall also form a consonant diad. Thinking in terms of ratios, we have available

PLATE XI

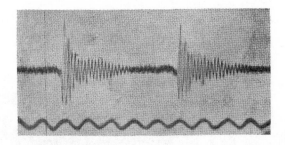

FIG. 9.2.—Record obtained by blowing excised larynx of a calf

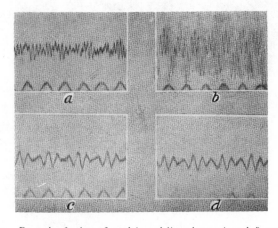

FIG. 9.3.—Records of voices of good (*a* and *b*) and poor (*c* and *d*) quality

PLATE XII

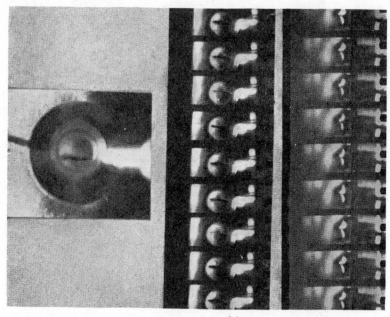

(a) (b)
FIG. 9.9.—Lip vibration at low frequency: (a) front view; (b) side view
(Martin, D. W., *Journ. Acous. Soc. Amer.*, Vol. 13, p. 305 (1942))

(*a*) the Minor Third, $6/5$; (*b*) the Major Third, $5/4$; (*c*) the Fourth, $4/3$; (*d*) the Fifth, $3/2$; (*e*) the Minor Sixth, $8/5$; (*f*) the Major Sixth, $5/3$. Remembering that to add intervals we multiply ratios, the possibilities within the octave are:

(1)	Two Minor Thirds	$= 6/5 \times 6/5 = \frac{36}{25}$		
(2)	Minor Third + Major Third	$= 6/5 \times 5/4 = \frac{3}{2}$	= Fifth	
(3)	Minor Third + Fourth	$= 6/5 \times 4/3 = \frac{8}{5}$	= Minor Sixth	
(4)	Minor Third + Fifth	$= 6/5 \times 3/2 = \frac{9}{5}$		
(5)	Minor Third + Minor Sixth	$= 6/5 \times 8/5 = \frac{48}{25}$		
(6)	Two Major Thirds	$= 5/4 \times 5/4 = \frac{25}{16}$		
(7)	Major Third + Fourth	$= 5/4 \times 4/3 = \frac{5}{3}$	= Major Sixth	
(8)	Major Third + Fifth	$= 5/4 \times 3/2 = \frac{15}{8}$		
(9)	Two Fourths	$= 4/3 \times 4/3 = \frac{16}{9}$		

Thus there are three possible combinations, (2), (3), and (7), and as either of the diads may be either above or below, there are six possible chords. Starting with *c* in each case, we may write them down as: (2) *c eb g*, *c e g*, (3) *c eb ab*, *c f ab*, (7) *c e a*, *c f a*. The same series of chords can be developed from the major and minor triads by inversion. Thus the chord *c e g* is called the common chord or the major triad in its root or fundamental position. Its first inversion is obtained by taking the octave of *c*, and we have *e g c*—it is the major chord of the sixth and third, the same as *c eb ab*. The second inversion is got by taking *e* up an octave, and we have *g c' e'*, the same as *c f a*—this is the major chord of the sixth and fourth. In the same way we can start with the minor chord in its root position, *c eb g*. Its first inversion is *eb g c'*, which is the same as *c e a*, and is the minor chord of the sixth and third, while its second inversion is *g c' e'b*, the same as *c f ab*, and is the minor chord of the sixth and fourth.

So far as beats between partials are concerned, the comparison of these triads introduces no new principle. But by considering the beats of partials alone it would be impossible to discriminate between the major and minor triads in their fundamental position. Both consist of a major third and a minor third, the only difference being that in the case of the major triad the major third is in the lower position, while in the minor triad the minor third occupies this position. We must therefore consider what other factors may possibly be involved, and one of these is the existence of combination tones.

Combination Tones and their Effect on Harmony.—When two notes are sounded together we are dealing with a much more complex phenomenon than we usually stop to imagine. Each note may be carrying with it anything up to twenty partial tones. And each partial tone of either series may make with each partial of its own or the other series the whole system of combination tones indicated on p. 31. When three or four notes are simultaneously sounded, the imagination is baffled. While it is neither possible nor desirable at present to take account of all this complexity, it is of interest to consider at least the first-order difference tone, which is very strong, and the first-order summational tone for some of the simple intervals.

Taking first the octave, we may assume that it is the interval c–c', the frequencies being 132 and 264. The first-order difference tone has a frequency $264 - 132 = 132$, which is the same as the lower note of the octave, and the first-order summation tone has a frequency $264 + 132 = 396$. This makes with c' an interval $\frac{396}{264} = \frac{3}{2}$, or a fifth. The summation tone is therefore g'. Either of these combination tones may form with either of the primary tones a second-order combination tone. The possibilities therefore are: (a) first-order difference tone and lower primary $132 + 132 = 264$, $132 - 132 = 0$; (b) first-order difference tone and upper primary $132 + 264 = 396$, $264 - 132 = 132$; (c) first-order summation tone and lower primary $396 + 132 = 528$, $396 - 132 = 264$; (d) first-order summation tone and upper primary $396 + 264 = 660$, $396 - 264 = 132$. Here the only new tones are 528 and 660. We note that $\frac{528}{264} = 2$, so that the note 528 is c''. We also note that $\frac{660}{528} = \frac{5}{4}$, so that the note 660 is a major third above c''—i.e., it is e''. The frequencies of second-order tones are therefore 132, 264, 396, 528, and 660. Considering the possible third-order tones in the same way, we find them to be 132, 264, 396, 528, 660, 792, and 924. Helmholtz found that third-order difference tones could be heard where partials were strong, but that even the second-order summation tones were almost impossible to hear, and if we accept this limitation then for the octave we have:

Primaries	132, 264
First-order tones . .	132, 396
Second-order difference tones .	132, 264
Third-order difference tones .	132, 264

Thus the only new note added is the first-order summation tone, 396. Considering the fifth in the same way, we may take the frequencies to be 132 (c) and 198 (g). In this case we have:

Primary tones 132, 198
First-order difference tone . . 66
First-order summation tone . 330
Second-order difference tones . $132 - 66 = 66$, $198 - 66 = 132$

Here the new notes are 66 and 330. We have $\frac{132}{66} = 2$, and therefore 66 is the octave below c, represented by C; and $\frac{330}{198} = \frac{5}{3}$, so that frequency 330 is a major sixth above frequency 198—i.e., it is a major sixth above g. It is therefore e'.

We can, of course, work by relative frequencies instead of actual frequencies. Thus, taking the interval of the fourth, the relative frequencies are $4 : 3$. The first-order difference tone is 1 and the first-order summation tone 7. The second-order difference tones are $4 - 1 = 3$ and $3 - 1 = 2$. The third-order difference tones are $4 - 3 = 1, 4 - 2 = 2, 3 - 3 = 0, 3 - 2 = 1$. Thus the new tones added by the combination have frequencies 1, 7, and 2 on the scale on which the primaries have frequencies 4 and 3. 2 is obviously an octave below 4—i.e., an octave below the upper primary—and 1 is an octave below this again—i.e., two octaves below the upper primary. 7 is not on our scale. It comes between the major sixth above 4—i.e., $\frac{5}{3} \times 4 = 6\frac{2}{3}$— and the minor seventh above 4—i.e., $\frac{16}{9} \times 4 = 7\frac{1}{9}$. Thus, if the fourth is c-f, the combination tones are F, F_1, and $e'\flat$. The series of combination tones for the various intervals is set out in the following table. In each column only the new tones introduced are inserted.

Interval.	Pri-maries.		Fre-quencies.	First-order tones.	Second-order difference tones.	Third-order difference tones.
Octave	c	c'	132, 264	396 (g')		
Fifth	c	g	132, 198	66 (C) 330 (e')	88 (F)	
Fourth	c	f	132, 176	44 (F_1) 308 (e') *	44 (F_1)	176 (f)
Major Sixth	c	a	132, 220	88 (F) 352 (f')	44 (F_1)	66 (C)
Major Third	c	e	132, 165	33 (C_1) 297 (d')	99 (G)	66 (C)
Minor Sixth	e	c'	165, 264	99 (G) 429 (a') *	66 (C)	198 (g)
Minor Third	e	g	165, 198	33 (C_1) 363 ($f'\sharp$) *	132 (c)	66 (C)

* Not exact.

The same analysis can be set out in staff notation as is done below, using minims for the primaries, crotchets for the first-order tones, quavers for the second-order difference tones, semi-quavers for the third-order difference tones. The summation tone is always recognizable as being higher in pitch than either of the primaries.

It will be seen that for the octave and fifth even the summation tone is a partial of one of the primaries, and so introduces no fresh element into the complex. For the other intervals that is not so. The summation tone is a foreign element, and this is an additional reason for the superiority of the octave and the fifth. In no case do the difference tones introduce any frequency which forms a dissonance with one of the primaries or with another difference tone.

Helmholtz has discussed at some length the relative harmoniousness of the major and minor triads. He assumes that the important tones are the three fundamentals forming the triad and the difference tones between them, taken in pairs, and between the fundamentals and their second partials taken in pairs. Thus, if the relative frequencies are given by $4 : 5 : 6$, then the relative frequencies of the second partials are $8 : 10 : 12$. On this scale we have the difference tones:

$$5 - 4 = 6 - 5 = 1 \text{ and } 6 - 4 = 2$$

we also have:

$$8 - 5 = 3 \qquad 8 - 6 = 2$$
$$10 - 4 = 6 \qquad 10 - 6 = 4$$
$$12 - 4 = 8 \qquad 12 - 5 = 7$$

By actual listening, Helmholtz arrived at the view that the tones 8 and 7 were relatively feeble, and that, in general, difference tones produced by primes widely separated in pitch contributed very little to the resulting impression produced by the chord. He therefore represents the primes 4, 5, 6 by minims, the tones 1, 2 by crotchets, the tones 2, 3 by quavers, and the tones 7, 8 by semiquavers. The tone 7 is a flattened *bb*, and is represented as such by a descending stroke in front. Setting out the three positions of the major triad in this way, and comparing it with the minor triad he obtains:

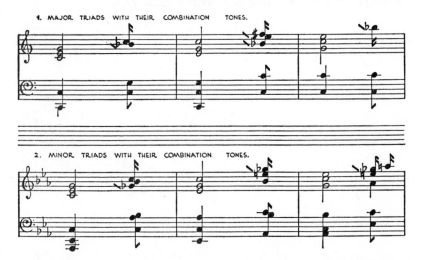

Commenting on this analysis, he remarks that in the major triads the difference tones between the primaries, and even the deeper difference tones between the second partials and primaries, are merely doubles of the tones of the triad in deeper octaves. For the minor triads, on the other hand, the difference tones between the primaries begin to disturb the harmonious effect. They are not near enough to beat, but they do not belong to the harmony. The lower difference tones formed by second partials and primaries (and represented by quavers) will in some cases actually beat with the primaries themselves.

For a further discussion of triads and tetrads the reader is referred to the original work of Helmholtz. It is of great interest as a piece of pioneer work, although the practical importance of some of it may now be doubted. The overtone structure of notes on the various instruments varies much more widely than Helmholtz supposed. The discussion applies strictly to true intervals correctly tuned or taken, whereas what we usually hear are tempered intervals incorrectly tuned or taken, and this has an important bearing on the matter of smoothness and harmoniousness. Further advance can only follow on further work directed to the relative importance of difference and summation tones, of combination tones of primaries with upper partials and upper partials with one another, of first- and second-order combination tones, &c. There is a rich field here for workers with this interest and with critical ears.

CHAPTER 11

SCALES AND TEMPERAMENT

Scales.—In the process of the development of music the first step was to select from the infinite variety of notes available the limited series to be used. The series of notes so selected is called a scale. The smallest interval in our own scale is a semitone. Between the two notes forming the interval lie at least twenty-five other distinguishable notes, none of which is used. Nor is this limitation a mechanical one due to the difficulties of the design of keyed instruments, for the use of a limited scale occurs where music has been developed by the human voice without any such hindrances. It is due to a necessity in the nature of music itself—the necessity that the movement in pitch which constitutes melody should proceed by definite and appreciable steps, and not by indeterminate gliding, and that successive notes should stand to one another in a simple and easily appreciated relationship. Only in this way can music make its appeal to the mind and to the emotions.

It must not be thought, however, that men first made a scale and then proceeded to use it for musical purposes. As Sir Hubert Parry has reminded us in his *Art of Music*, 'Scales are made in the process of endeavouring to make music and continue to be altered and modified generation after generation, even till the art has arrived at a high degree of maturity'. It is not surprising, then, that though all nations use scales, the scales themselves differ greatly. There is no standard form imposed by nature. On the other hand, there are marked similarities between the different scales which point to some cause inherent in nature which it is the business of the physicist to explore. Thus everywhere the octave is taken as a kind of unit of the scale and divided into degrees by other notes. If we take our own C major scale, the octave is divided into seven degrees by the notes corresponding to the white keys. Although our own diatonic scale is a comparatively late development, owing its particular structure to the rise of harmony, and therefore subject to factors which have played no part in the scales developed for homophonic or one-part music, it is worth while thinking in terms of our own scale and using the nomenclature adapted to it. We may remind ourselves that the notes have the following names:

 c tonic
 d second or supertonic
 e third or mediant
 f fourth or subdominant
 g fifth or dominant
 a sixth or superdominant
 b seventh or subtonic or leading note
 c′ octave.

The ordinal number of a note of the scale is also the name given to the interval which that note forms with the tonic.

There is, however, another scale in common use with C as the keynote or tonic—that of C minor. In its commonest form it is

c tonic	*g* fifth or dominant	
d second or supertonic	*a♭* sixth or superdominant	
e♭ third or mediant	*b♭* seventh or subtonic	
f fourth or subdominant	*c′* octave.	

The intervals from the tonic to the fourth and to the fifth respectively in both scales being the same, these intervals are known as a perfect fourth and a perfect fifth. On the other hand, we have the major third *c–e* and the minor third *c–e♭*; the major sixth *c–a* and the minor sixth *c–a♭*; the major seventh *c–b* and the minor seventh *c–b♭*. The interval *c–d* is a major second or tone, the interval *b–c′* is a minor second or diatonic semitone.

When the octave is divided into two intervals, each is said to be the inversion of the other.   Thus we have for the major scale:

Major Second	Perfect Fourth	Major Sixth
Minor Seventh	Perfect Fifth	Minor Third
Major Third	Perfect Fifth	Major Seventh
Minor Sixth	Perfect Fourth	Minor Second

and for the minor scale:

Major Second	Perfect Fourth	Minor Sixth
Minor Seventh	Perfect Fifth	Major Third
Minor Third	Perfect Fifth	Minor Seventh
Major Sixth	Perfect Fourth	Major Second

The origin of our own major and minor scales can be traced with fair certainty to the music of the Greeks. Music undoubtedly played an important part in the life of the Greek people. Plato assigned to it a prominent role in education, maintaining that it was effective in producing a certain inner harmony which other subjects of education failed to give. It is said that all Greek citizens had some training in music and were able to take part in the music which accompanied public functions. Unfortunately the actual music has been almost completely lost, only a few fragments survive. The Greek contributions to the *theory* of music, on the other hand, have, on the whole, been well preserved in a mass of writings, especially those coming from the followers of Pythagoras. The records of musical culture show that there was considerable development as early as 1200 B.C. The instrument most closely associated with the development of the scale was the tetrachord, an instrument of four strings, as its name suggests. The earliest tuning seems to have been that of the ancient tetrachord of Olympus, which may be thought of in terms of the white notes of the piano, *a*, *f*, *e*, or *e*, *c*, B—i.e., a downward fourth with an ornamental third. Taking *e*, *c*, B, and adding the minor third below *e*, we have *e*, *c♯*, *c*, B, which was the chromatic tetrachord, or, taking the minor third up from B, we have *e*, *d*, *c*, B, the diatonic tetrachord. It will be noticed that in the two cases the fourth is differently subdivided. Thinking of the scale as ascending from B to *e*, the diatonic tetrachord has semitone, tone, tone, and the chromatic tetrachord semitone, semitone, minor third.

The great variety of scales, or 'modes' as they are called, is indicated by the following table, in which T stands for a tone of approximately the same value as our tone, while *t* stands for a semitone whose value is approximately the same as our semitone:

Species		Dorian.			Phrygian.			Lydian.		
	Diatonic . .	*t*	T	T	T	*t*	T	T	T	*t*
	Chromatic .	*t*	*t*	T + *t*	T + *t*	*t*	*t*	*t*	T + *t*	*t*
	Enharmonic .	$\frac{t}{2}$	$\frac{t}{2}$	2T	2T	$\frac{t}{2}$	$\frac{t}{2}$	$\frac{t}{2}$	2T	$\frac{t}{2}$

These recognized modes give us nine different ways of arranging the intervals of the tetrachord. The popularity of the enharmonic species seems to have been short-lived. The scale was extended by adding a second tetrachord to the first. Sometimes the first string of this second tetrachord was tuned to the fourth string of the first, but later there was a 'disjunctive' tone between the two. Taking the disjunctive arrangement and the diatonic species, we have three modes:

Dorian	.	.	.	*t*	T	T	T	*t*	T	T
Phrygian	.	.	:	T	*t*	T	T	T	*t*	T
Lydian	.	.	.	T	T	*t*	T	T	T	*t*

These three modes can be reproduced fairly nearly by using the white notes on the piano as follows:

Dorian	.	.	*e*	*f*	*g*	*a*	*b*	*c'*	*d'*	*e'*
Phrygian	.	.	*d*	*e*	*f*	*g*	*a*	*b*	*c'*	*d'*
Lydian	.	.	*c*	*d*	*e*	*f*	*g*	*a*	*b*	*c'*

and we see that the Lydian coincides with our ordinary scale in C major, while the Dorian and Phrygian, containing no accidentals, seem also to be in the key of C major, but start and finish on the wrong note.

The Greek system may be said to have reached maturity when a compass of some two octaves was mapped out into a series of seven modes, fairly represented by playing scales on the white notes of the piano only, and starting successively with B, C, *d*, *e*, *f*, &c., thus:

Mixolydian	B	*c*	*d*	*e*	*f*	*g*	*a*	*b*				
Lydian		*c*	*d*	*e*	*f*	*g*	*a*	*b*	*c'*			
Phrygian			*d*	*e*	*f*	*g*	*a*	*b*	*c'*	*d'*		
Dorian				*e*	*f*	*g*	*a*	*b*	*c'*	*d'*	*e'*	
Hypolydian					*f*	*g*	*a*	*b*	*c'*	*d'*	*e'*	*f'*
Hypophrygian						*g*	*a*	*b*	*c'*	*d'*	*e'*	*f'*	*g'*
Hypodorian or Aeolian							*a*	*b*	*c'*	*d'*	*e'*	*f'*	*g'*	*a'*

Transposing the modes so that they all start with *c*, we have:

Greek name.									Ecclesiastical name.
Mixolydian . .	*c*	*d♭*	*e♭*	*f*	*g♭*	*a♭*	*b♭*	*c'*	Ionian
Lydian . . .	*c*	*d*	*e*	*f*	*g*	*a*	*b*	*c'*	Mixolydian
Phrygian . . .	*c*	*d*	*e♭*	*f*	*g*	*a*	*b♭*	*c'*	Dorian
Dorian . . .	*c*	*d♭*	*e♭*	*f*	*g*	*a♭*	*b♭*	*c'*	Aeolian
Hypolydian . .	*c*	*d*	*e*	*f♯*	*g*	*a*	*b*	*c'*	Phrygian
Hypophrygian . .	*c*	*d*	*e*	*f*	*g*	*a*	*b♭*	*c'*	Locrian
Hypodorian or Aeolian	*c*	*d*	*e♭*	*f*	*g*	*a♭*	*b♭*	*c'*	Lydian

In each case the scale is heptatonic—i.e., it consists of seven notes to the octave, and these seven notes divide the octave into seven degrees, of which five are larger, and are called tones, and two re smaller, and are called semitones. Variety in composition is obtained by using different modes, in each of which the tones and semitones are differently distributed. These modes are all produced in homophonic music—i.e., in melody only.

The Greek modes formed the basis of the musical system of the early Christian Church, and were codified and simplified by Bishop Ambrose of Milan in the fourth century. He recognized four authentic modes. The tonic of each was the lowest note of the scale, and the dominant was the fifth, except when this was *b*. Thus the authentic modes were:

Tonic.							Dominant.						
d	.	.	.	*e*	*f*	*g*		*a*	.	.	.	*b*	*c′*	*d′*
e	.	.	.	*f*	*g*	*a*	*b*	*c′*	.	.	.	*d′*	*e′*	
f	.	.	.	*g*	*a*	*b*		*c′*	.	.	.	*d′*	*e′*	*f′*
g	.	.	.	*a*	*b*	*c′*		*d′*	.	.	.	*e′*	*f′*	*g′*

To these Authentic Modes Pope Gregory added four ' Plagal ' Modes. These have the same key-note as the corresponding Authentic Mode, but the tonic, on which the composition must end, is placed a fourth above the lowest note which can be used, and which may be regarded as the first note of the scale. Thus we have:

			Tonic.				
A	B	*c*	*d*	*e*	*f*	*g*	*a*
B	*c*	*d*	*e*	*f*	*g*	*a*	*b*
c	*d*	*e*	*f*	*g*	*a*	*b*	*c′*
d	*e*	*f*	*g*	*a*	*b*	*c′*	*d′*

In the sixteenth century Glarean in his *Dodecachordon* (1547) recognized twelve modes, to which unfortunately he assigned the wrong Greek names. Some of these modes were purely theoretical and had never been used. Some had been used, but had fallen into disuse. With the introduction of harmony the situation hardened, and only two modes survived, the scale on the white notes beginning with *c*, the old Greek Lydian, which survives as our major scale, and the scale on the white notes starting with *a*, the old Greek Aeolian, which survives as its relative minor.

The minor scale involves a minor third from the tonic. There are three possible forms. Starting from *c* in each case these are:

(1)	*c*	*d*	*e♭*	*f*	*g*	*a*	*b*	*c′*
(2)	*c*	*d*	*e♭*	*f*	*g*	*a♭*	*b*	*c′*
(3)	*c*	*d*	*e♭*	*f*	*g*	*a♭*	*b♭*	*c′*

Of these (3) is the commonest, and is almost always used in moving downwards. (1) is sometimes used in moving upwards, and in the eighteenth century (2) was a common ascending form. It will be seen that (3) is the Hypodorian or Aeolian mode.

Other scales and modes have, of course, been developed by other peoples and nations. So long as music is homophonic, a very great variety is possible. Whether the development of harmony which has been mainly responsible for the disappearance of modes in Europe will spread to other parts of the world and produce like changes, it is impossible to say, but it seems not unlikely. The Chinese had originally a pentatonic scale represented by the notes f g a c' d'. They had semitones, but only for purposes of modulation. Their scale is the same as

$$f\sharp \quad g\sharp \quad a\sharp \quad c'\sharp \quad d'\sharp$$

i.e., the black notes of the piano. A later form of the scale used the same notes, but the mode started from d instead of from f. This pentatonic scale is found in American Indian music and also in Celtic music. A large number of Scottish melodies can be played on the black notes of the piano, and the large steps from the third to the fourth and from the fifth to the sixth notes of the scale are characteristic of Scottish music. In India a very complex system of scales was developed, involving half-tones and even quarter-tones. In Persia, too, a great variety of modes was used, and subsequently adopted and further modified by the Arabs. They appear in Spanish folk-music. Our ears are so accustomed to the major and minor diatonic scales that others do not readily appeal to us, and some effort on our part is necessary if we are to appreciate what is unfamiliar. But the making of new modes or scales has not ceased. In the work of Debussy we frequently find the whole tone scale represented by

$$c \quad d \quad e \quad f\sharp \quad g\sharp \quad a\sharp$$

This scale is also found in ancient oriental music. Scriabin used a scale represented by

$$c \quad d \quad e \quad f\sharp \quad a \quad b\flat$$

omitting the g, and this gives a distinctive character to his music. There are, no doubt, still fresh discoveries to be made in this realm.

Rise of Harmony.—Homophonic music—i.e., one-part music—is the original form, and still holds the field among the Chinese, Indians, Arabs, and Turks. But it is very limited in its possibilities. It can only be used for short pieces or for the accompaniment to poetry. The great developments which have taken place in European music have been due to the introduction first of polyphony and later of harmony. Polyphony, or many-part music, seems to have arisen first through the necessity of providing some means whereby boys and men, or men and women, could sing together. At first the melody was duplicated at the interval of an octave—a process called *magadizing*. Later the melody was duplicated at an interval of a fourth or a fifth—known as *organizing*. Later there began the practice of weaving different melodies together so that the parts were independent, and yet by minor adjustments avoiding harsh discords at any point. This was the principle of the discantus, popular in France and Flanders in the eleventh century. Helmholtz says that occasionally a sacred melody and a rather slippery popular song were joined in an unholy wedlock in this way.

But there is a marked difference between this kind of polyphony and harmony. So long as the music is thought of as a series of independent melodies woven together, we are in the realm of counterpoint. The movement of the parts is important. We think of the music ' horizontally '. In harmony the primary conception is a single chord, and we think of the music vertically. We must not press the distinction unduly, although it is a real one; the transition was gradual and the conception of harmony did not become established until the seventeenth century. It was greatly helped in its early stages by the Reformation, which resulted in a demand for simply harmonized chorales adapted for Protestant congregational singing. The reaction of harmony on the scale was twofold. In the first place, it fixed the pitch of the notes with greater precision. An inaccurate interval taken in melody may pass unnoticed, whereas the same inaccuracy in harmony would turn a smooth concord into a harsh discord. Then again it puts a premium on the scale offering the largest choice of concordant intervals.

Alongside of the development of harmony went the development of ' tonality '. This involved attaching unique importance to one note in the mode. There are indications in the writings of Aristotle that this principle was vaguely recognized, but his references are without independent support. In the ecclesiastical modes the principle was also recognized, but was still very indefinite. With the rise of harmony the principle acquired an altogether new importance. All other notes in the scale were related to the tonic, and in this way a new principle of unity was established. A few consecutive chords are sufficient to establish the key. The whole mass of tones and the consecutive harmonies must stand in a close and clearly perceptible relationship to the tonic, while the composition must be developed from it and finally return to it.

Once this principle had become established the musical variety made possible by a large number of modes ceased to be important, and variety was sought rather in the use of the same mode at a different level of pitch—i.e., by modulation into a new key. Thus the only two modes to survive—as has already been noted—have been the major and minor modes, which revealed themselves as more effective for harmonic development than any of the others.

Emotional Character of Different Keys.—The Greeks associated different emotions with their different modes. The Dorian mode played an important part in the education of Spartan boys. It was said to suggest dignity, manliness, self-dependence, and courage. The Phrygian mode was also regarded as inspiring. The Lydian mode, on the other hand, was supposed to conduce to softness and self-indulgence, and to share with the Ionian or Hypophrygian mode a certain voluptuous and orgiastic character. Plato would have forbidden the use of the last two modes in his ideal republic. If the character which is supposed to attach itself to a mode has any real basis, and is not due simply to the association of ideas—the conventional use of a particular mode for a particular type of music—then it must be related to the distribution of the semitones in the mode, because this is what constitutes the variety. Certainly most musicians would agree in ascribing entirely different emotions to the two surviving modes—our major and minor keys. The ' minor key ' has become an accepted figure of speech. Here the same explanation may be given. The character may be associated with the order in which the semitones occur in the scale, the distinction between the minor third and the major third or between the minor sixth and the major sixth. But can we associate any definite emotional character with different keys in the same mode? Is there any difference in character between music in C major and music in G major, or between C major and

C♯ major?   Musicians seem to have held that there is.   Berlioz, Schumann, and Beethoven can all be quoted in support.   Of Handel, Leichtentritt speaks as follows:

' As to tonality in Bach and Handel, we have of late gained new insight into a system of surprising extension, the existence of which was not even suspected twenty-five years ago. Handel, the great dramatist, makes a most scrupulous choice of keys for the arias in his operas and oratorios.   It matters very much to him whether he writes a piece in F major, or F sharp major, or F flat major, in F minor or F sharp minor. For him every one of these keys has well-defined colour, atmosphere, and meaning, to which he adheres strictly during his entire artistic career of over fifty years.   F major, for instance, is the key of the pastoral idyll all through the eighteenth century, and it is certainly not by chance that Beethoven a hundred years later chooses it for his Pastoral Symphony and for his " Spring " Sonata for violin and piano, op. 24.   F sharp major for Handel is what one might call a transcendental key;   indeed, all keys with signatures of five, six, seven and even eight and nine sharps are associated by him with the idea of heaven, with ecstatic visions of a world beyond earthly toil and pain, with eternal peace and heavenly consolation.   F minor and F sharp minor are both tragic keys, but there is a subtle distinction between them.   F minor is generally chosen for the expression of profound sadness and melancholy;   it is a dark, pathetic, lamenting key.   Certain aesthetic conceptions of the eighteenth century still survive in Beethoven, and again it is not by chance that he wrote his " Egmont " Overture in F minor, and that the gloomy prison scene in the second act of his " Fidelio " has the F minor tonality.   For Handel F sharp minor is full of tragic intensity, less melancholy and sentimental than F minor. It sometimes has a heroic note, a sound of brave resistance to a cruel fate.   G minor in Handel's operatic music is preferably used for the agitation of jealousy;   E minor is reserved for the expression of an elegiac mood.   It is interesting to remember that one of the most admired masterpieces of elegiac music, Brahms's Fourth Symphony, chooses the same key, E minor;   we also remember that Brahms was a close student of Handel and took many a hint from this great master. G major in Handel suggests bright daylight, sunshine, green meadows. C major is used to express manly vigour, military discipline and the elemental power of nature. Beethoven uses it in the same way. It is the key for plain, straightforward action, without psychological complication; it is the *Naturtonart*, as the Germans call it. In this way one might go through all the other keys. Handel's entire harmonic system and style of modulation is based on the underlying meaning of the various keys.'

On the other hand, Schubert may be cited against this view. It is said of him: [1]

> ' Schubert, one of the most spontaneous composers who ever lived, had so little instinctive conviction concerning the atmosphere created by, and pertaining to, a particular key that he not infrequently transposed his songs, without modifying either the vocal line or the pianoforte accompaniment. He did this when assembling some of his finest songs for inclusion in the Winterreise cycle. The original version of "Mut" was written in A minor and the version finally published was in G minor without further alteration.'

Helmholtz gives a somewhat tentative support to the view that there are distinctive characteristics of the various keys, and suggests possible explanations. He points out that whatever may have been true of the old modes, the introduction of equal temperament has completely obliterated all differences except the difference of pitch between one major key and another. He therefore abandons the possibility that there can be any difference in character between the different keys on an instrument like the organ. But in the case of the pianoforte he points out that the black keys and the white keys set the hammers in action by levers of different length, and the number of black notes in the scale and their position may influence the character of the key. In a similar way, in the case of stringed instruments, he suggests that the position in the scale of the notes produced on the open string may be a determining factor. It is hard for the physicist to accept these slight differences as a basis for differences of emotional quality, and the difficulty is increased when it is claimed that the key of F sharp major is different in character from that of G flat major, although in performing any piece in these two keys exactly the same notes on the pianoforte are struck. Nor among those who claim that particular keys have distinctive characters is there any real unanimity as to the characters to be assigned to them. It therefore seems wiser to assume, without being too dogmatic, that the association of particular keys with music of a particular type, and especially with familiar examples, has given rise to a belief in distinctive emotional characters for which there is in fact no rational foundation, although the possible association with absolute pitch (p. 58) should not be entirely discounted.

Scientific Basis of the Scale.—We have seen that in all scales the interval of the octave is fundamental, and the fifth and the fourth almost, if not quite, universal. It is tempting to regard that fact as due to the harmonious nature of these intervals. But at once we are reminded of the fact that scales were modified by harmony, but existed long before harmony was developed in Europe, and still exist where harmony has never been developed. Primitive people singing in a cave would find one note running into the next because of reverberation, and in some very reverberant buildings one can sing four successive notes and hear them as a chord. In the case of vibrating strings also, where the damping is not great one string may still be vibrating when the next is plucked or struck. There is no need, however, to depend on harmony for an explanation of the more obvious musical intervals. In the case of an octave the upper note contains every even partial of the lower, and the possession of these partials in common may well account for the simple relationship which the ear appreciates when the two notes are sounded successively. A similar relationship holds for the fifth and the fourth, as has already been shown (p. 161).

The earliest contributions to the development of the theory of the musical scale come from the school of philosophers founded by Pythagoras, a school of great influence and importance. No authentic original works by Pythagoras himself have survived, but his work has become known through the writings of his disciples. He believed that numbers were the ultimate explanation of all things. He found that simple musical intervals were given by the notes produced by the two segments of a stretched string if the point of division gave segments whose lengths were in a simple numerical ratio (see p. 50). The simplicity of this ratio was for him the ultimate *explanation* of the pleasant nature of the interval. He concluded that 'intervals in music are rather to be judged intellectually through numbers than sensibly through the ear' and 'the simpler the ratio of the two parts into which the vibrating string is divided the more perfect is the consonance of the two sounds'.

We push his explanation a stage farther back when we take account of the relation between vibrating length and frequency, and realize that if the vibrating lengths are in a simple ratio, so are the frequencies (in the inverse ratio), and therefore the two notes must possess partials in common which can explain the relationship which the ear apprehends. Pythagoras proceeded to derive a chromatic scale based entirely on octaves and fifths— the two simplest and most obvious intervals. He knew that the octave was produced by two segments of a string whose lengths are in the ratio 2 : 1, and the fifth by two segments whose lengths are in the ratio 3 : 2. Using this and the ratio of lengths for the octave (2 : 1), Pythagoras worked out the ratio of lengths of strings for all the notes of a complete chromatic scale. Knowing that the frequencies of vibration are inversely proportional to the vibrating lengths, we can forget about the lengths, and think in terms of frequencies.

We shall appreciate the method of Pythagoras best, perhaps, if we follow his method of deriving the scale but measure all the intervals of the scale in cents (see p. 53). We may remind ourselves of the size of this unit by noting that there are 100 cents to the tempered semitone of the pianoforte and 200 cents to the tempered tone. The octave contains 1,200 cents and the perfect fifth contains 702 cents. To go up an octave, therefore, we add 1,200 cents, and to go up a fifth we add 702 cents. Conversely, to descend the scale we subtract. In deriving the scale it is simplest to think in terms of the scale of C major, the key of the white notes of the pianoforte, and to name the notes by the letters so that the intervals are

	Second	Third	Fourth	Fifth	Sixth	Seventh	Octave
c	d	e	f	g	a	b	c'

Starting with c, the fifth is g, 702 cents above c. Another fifth brings us to d', 1,404 cents above c. Bringing this down an octave to d we have 1,404 − 1,200 or 204 cents above c. Successive fifths calculated in the same way give a, e, and b. Taking a fifth below *c* and bringing it up or taking a fifth down from *c'*, we get *f*. The intervals from *c* of the various notes in this scale of seven notes, measured in cents, are as follows:

Interval.				Note.	Cents above *c*.
Second	.	.	.	*d*	204
Third	.	.	.	*e*	408
Fourth	.	.	.	*f*	498
Fifth .	.	.	.	*g*	702
Sixth .	.	.	.	*a*	906
Seventh	.	.	.	*b*	1,110
Octave	.	.	.	*c*¹	1,200

Looking at the intervals between consecutive notes of the scale, we see that they are of two sizes—204 cents (*c–d*, *d–e*, *f–g*, *g–a*, *a–b*) and 90 cents (*e–f*, *b–c'*). The larger of these intervals is the Pythagorean tone, and the smaller the Pythagorean diatonic semitone or limma. Here, then, we have a scale of seven notes to the octave which strongly resembles our major scale. It has five tones and two semitones, and these are distributed in the order tone, tone, semitone, tone, tone, tone, semitone, just as in our scale.

What happens if we pursue further this series of fifths? The last note reached by our ascending fifths was *b*. The note we reach by going up a fifth and coming down an octave gives us 1,110 + 702 − 1,200 = 612 cents (above *c*), and lies between *f* and *g*. If we call it *f* sharp, then it lies 612 − 498 = 114 cents above *f*, and we have a new interval which is called the Pythagorean chromatic semitone. We can obviously make this new note, *f* sharp, the starting point of a new series of fifths which gives us in turn *c* sharp, *g* sharp, *d* sharp, *a* sharp, *e* sharp, *b* sharp—i.e., each note of the scale will have its corresponding sharp 114 cents above. Notice that the chromatic semitone which separates the natural note from its sharp is not the same as the diatonic semitone (90 cents) which occurs in the scale. Also neither of them is exactly half a tone. The chromatic is greater than half a tone, the diatonic is less, but the two together make up a tone. The difference between the two semitones is a Pythagorean comma, and is 24 cents. A more exact value for the fifth would have given us 23·46 cents. It is the interval got by ascending twelve fifths and descending seven octaves. Observe that on this scale every note has its sharp, and e sharp is not the same as f, nor is b sharp the same as c.

If we had continued with the series of fifths, we should have come through b sharp to f double sharp, two chromatic semitones above f. This note is $2 \times 114 = 228$ cents above f, and therefore $228 - 204 = 24$ cents or one Pythagorean comma above g. All the notes obtained by proceedings further with an ascending series of fifths would differ by one comma from notes already obtained, and would not be sufficiently different to be useful for melody.

If instead of ascending fifths we take descending fifths, the first, as we saw, gives us f. Taking a fifth down from f and raising it an octave, or what comes to the same thing, going up an octave and down a fifth, we have a note which makes with c an interval $498 + 1{,}200 - 702 = 996$ cents. It comes between a and b, so we call it b flat. It is 114 cents—i.e., one chromatic semitone below b. Successive steps in this downward series of fifths give us e flat, a flat, d flat, g flat, c flat, f flat. One more step brings us to b double flat, which lies fairly close to a. Here, then, we have a quite elaborate scale worked out completely by considering only the two simplest intervals—the octave and the fifth. It resembles closely our own scale. Its basis is five tones (all equal) and two semitones to the octave. It differs in having a sharp and a flat for each of its seven notes, making twenty-one in all, instead of the arrangement of our scale, whereby the sharp corresponding to one note is used also as the flat for the note next above it, so that we are satisfied with twelve notes in all. We shall make a more exact comparison later on. Meantime it is worth noting the distribution of sharps and flats round one of the semitones of the scale. Thus we have:

Note	d	$e\flat$	$d\sharp$	$f\flat$	e	f	$e\sharp$	$g\flat$	$f\sharp$
Interval above c	204	294	314	384	408	498	522	588	612

A somewhat different theoretical scale was developed by Aristoxenus and his successors. He lived in the third and fourth centuries B.C., and was a musician and the son of a musician. Much of his work has been lost, but there still survives a treatise on Harmonics in three volumes. He is reputed to have laid more stress on the aesthetic appreciation of music, which it is the business of mathematics to explain, and less stress on a mathematical theory, to which aesthetic appreciation was expected to conform.

The Greek Scale, as we have seen, was built up on a series of tetrachords. Each tetrachord covered the interval of a fourth. Aristoxenus based his scale on two tetrachords with a disjunctive tone between (p. 174). The highest note of the upper tetrachord was a perfect fifth above the highest note of the lower tetrachord, and each tetrachord embraced the interval of a fourth. This means that the disjunctive tone was the difference between the perfect fifth (702 cents) and the perfect fourth 498 cents) or 204 cents, the same as the Pythagorean tone. The particular tuning of the tetrachord was that in which, if we assume the four notes to be $c\ d\ e\ f$, d was a Pythagorean tone above c, e a major third above c, and f a perfect fourth above c. Measured in cents, we already know the Pythagorean tone to be 204 cents, and the perfect fourth to be the difference between the octave and the perfect fifth—i.e., $1{,}200 - 702 = 498$ cents. The major third (as we now know it) is the interval between two notes whose frequencies are in the ratio $5:4$. In cents it is 386.4 cents. The intervals then work out as follows:

c	d	e	f
0	204	386	498

If we add a second exactly similar group of four, starting with g, a perfect fifth above c, we have:

	c	d	e	f	g	a	b	c'
Cents above c .	0	204	386	498	702	906	1,088	1,200

In this scale, then, the intervals are not all the same as in the Pythagorean scale. The second, fourth, fifth, and sixth are the same, but the third and seventh are different. Also when we examine the degrees of the scale we see that they are in succession (in cents) 204, 182, 112, 204, 204, 182, 112. Thus, as in the Pythagorean scale, we have tone, tone, semitone, tone, tone, tone, semitone. We notice, however, that the tones are of two different sizes, the minor or lesser tone (182) and the major or larger tone (204), while the semitone is greater than the half of either of them. It is less than the chromatic semitone, but greater than the diatonic semitone, of Pythagoras. This scale was later modified by Zarlino, Maître de Chapelle at St. Mark's, Venice, about 1560, on the ground that the association of two identically tuned groups of four notes gave a certain monotony. The change made was to interchange the larger and lesser tones in the second group. This meant slightly lowering the *a* so as to be only 182 cents above *g*. The intervals from *c* now were:

	c	d	e	f	g	a	b	c'
Cents above c	0	204	386	498	702	884	1,088	1,200

The sixth now differs from that of the Pythagorean scale. In order that we may see the relationship of the notes in this scale and that of Pythagoras from another point of view, we may translate back from cents into frequency ratios, and we find the following:

	c	d	e	f	g	a	b	c'
Zarlino	1	9/8	5/4	4/3	3/2	5/3	15/8	2
Pythagoras	1	9/8	81/64	4/3	3/2	27/16	243/128	2

Thus, although the Pythagorean scale is derived very simply by using only the octave and the fifth, it produces some very complicated fractions, especially for the intervals of the third, the sixth, and the seventh. Now, the greater simplicity of the fractions in the scale of Zarlino has an importance which is not merely arithmetical. Arithmetical simplicity in the ratios of the frequencies carries with it the possession of low-order partials in common, and therefore indicates a relationship which the ear recognizes and appreciates. The construction of the scale of Pythagoras based on fifths and octaves affords no intervals except the fifth and its inversion, the fourth. It offers four of each within the compass of the octave. The scale of Zarlino, on the other hand, offers two minor thirds, three major thirds, four perfect fourths, three perfect fifths, one minor sixth, and two major sixths. In addition to this, it fits in with the principle of tonality. Four notes of the scale form simple intervals with the key note—the third, fourth, fifth, and sixth. The remaining two cannot be fitted in so as to be simply related to the tonic, but they can be, and are, fitted in so as to be simply related to one of its nearest relatives—the dominant. The second note of the scale is a perfect fourth below it, and the seventh of the scale is a major third above it.

Most important of all, notes which stand to one another in a simple frequency ratio form a concord when sounded together. The rise of harmony has therefore enhanced the demand for pairs of notes satisfying this condition, and all the intervals already named form concords. Because it is built up in this way, the scale we have just considered is usually called the natural or true scale.

The difference between this scale and the scale of Pythagoras can be seen from the following table:

Interval .	Second	Third	Fourth	Fifth	Sixth	Seventh	Eighth
Pythagorean .	204	408	498	702	906	1,110	1,200
Natural or True.	204	386	498	702	884	1,088	1,200

Thus the second, fourth, fifth, and octave are all identical, while the Pythagorean third, sixth, and seventh are each 22 cents sharp. This interval is the comma of Didymus, generally denoted by the term comma when unqualified. It is, in terms of ratios, the difference between the Pythagorean third $(81/64)$ and the true major third $(5/4)$—i.e., $\frac{81}{64} \div \frac{5}{4} = \frac{81}{80}$.

Returning to our method of justifying this scale by the simple relationship of its notes to the key-note, we find that a direct relationship gives us e, f, g, and a. The direct relationship of d and b is too remote to have any practical importance. This leaves us with two large gaps, one between c and e, and one between a and c', which we filled by taking the notes d and b, both simply related to the dominant g. We might, however, have looked for notes related to c', which is even more closely related to c than g is. This would not have helped us to fill our gaps, but it would have given us alternative notes for a and e. The major third below c is $1,200 - 386 = 814$ cents. This falls between g and a, and is called a flat. It is 70 cents below a, and the interval is called a chromatic semitone. Similarly a major sixth below c' is $1,200 - 884 = 316$ cents. This is e flat, a chromatic semitone below e. If now we take notes related to the sub-dominant, f, we find in the upper gap the note one fourth above f $(498 + 498 = 996$ cents). This comes between a and b, and is 92 cents below b. We call it b flat, and it is a semitone (70) plus a comma (22) below b. Thus we have added three new notes to the octave: a flat, e flat, and b flat. Substituting these notes for a, e, and b, we have the scale

	c	d	e♭	f	g	a♭	b♭	c'
Cents above c .	0	204	316	498	702	814	996	1,200

These are the notes required for the minor mode in its descending form. The minor mode in its ascending form requires only the *e* flat as an additional note to the original eight, and the mode can, of course, be realized (except for comma differences) by starting with A as our key note without any new notes at all so far as the descending form is concerned. The ascending form would require *f* sharp and *g* sharp, for reasons already given. If, then, we are satisfied with the use of two keys, C major and its relative minor, A minor, we require only nine notes to the octave. If we use instead C major and C minor, we shall require ten notes. What happens if we wish to use other keys?

The Need for Temperament.—Music which remains always in one key gives a sense of monotony which is quite intolerable, and this is avoided by frequent ' modulation ' or change of key, followed sooner or later by a return to the key of the composition. The commonest modulation is that from the key of the tonic into the key of the dominant—e.g., from the key of C major to the key of G major. The first step in the major mode is a major or larger tone. But *a*, which is the first step from *g*, is the sixth from *c*, and is separated from *g* by a minor or lesser tone. Thus to make possible the major mode starting from *g*, we must have a new *a* which is a comma sharper. Then again, the interval from *b* to *c'* is a diatonic semitone, so that the interval from *f* to *g* ought to be a diatonic semitone. Here, then, we require a new note which is a diatonic semitone below *g*. We call it *f* sharp. The other notes of the key of C major will serve also for the key of G major, as will be seen if we set them out in cents:

	c	*d*	*e*	*f*	*f♯*	*g*	*a*	*b*
Key of C major .	0	204	386	498	—	702	884	1,088
Key of G major .	0	204	386	—	590	702	906	1,088

F sharp is the seventh of the scale of G major, and is therefore a diatonic semitone below *g*. The chromatic semitone between *f* and *f* sharp is thus the difference between the tone *f–g* and the diatonic semitone *f♯–g*. These intervals in ratios and savarts are:

	Ratio.	Cents.
Tone *f–g*	$\frac{9}{8}$	204·0
Diatonic semitone *f♯–g* . .	$\frac{16}{15}$	111·7
Chromatic semitone *f–f♯* . .	$\frac{9}{8} \times \frac{15}{16} = \frac{135}{128}$	92·2

In this case the tone is the larger tone. In the case of the scale of E major, however, the seventh is *d* sharp, and the tone *d–e* is the lesser tone corresponding to the ratio $\frac{10}{9}$. In this case we have:

	Ratio.	Cents.
Tone d–e	$\frac{10}{9}$	182·4
Diatonic semitone $d\sharp$–e . .	$\frac{16}{15}$	111·7
Chromatic semitone d–$d\sharp$ . .	$\frac{10}{9} \times \frac{16}{15} = \frac{25}{24}$	70·7

The difference in the two chromatic semitones is, of course, the same as the difference in the two tones—namely, the comma $\frac{81}{80}$ or 21·5 cents.

Returning to a consideration of the change of key, a modulation from the key of G major into the key of its dominant would take us into the key of D major, and would require a new e sharper by a comma, and a new c which would differ by a chromatic semitone from c, and would be c sharp. We can set out our table of keys thus:

Key.	c	$c\sharp$	d	$d\sharp$	e	f	$f\sharp$	g	$g\sharp$	a	$a\sharp$	$b\flat$	b
C . .	0		204		386·4	498		702		884·4			1,088
G . .	0		204		386·4		590·4	702		906			1,088
D .		92·4	204		408		590·4	702		906			1,088
A .		70·8	182·4		386·4		568·8		772·8	884·4			1,088
E .		70·8		274·8	386·4		590·4		772·8	884·4			1,088
B .		92·4		274·8	386·4		590·5		772·8		976·8		1,088
F .	0		182·4		386·4	498		702		884·4		996	

It will be noticed that in the case of $c\sharp$, d, e, $f\sharp$, a, we require two notes separated by the comma 21·5 cents. The interval between $a\sharp$ and $b\flat$ is 19·2 cents. It is the difference between two diatonic semitones (223 cents) and one larger tone (204 cents).

Note that since all values are rounded to one decimal place, small discrepancies may appear.

If, then, we are to be free to use all these keys, we shall require eighteen notes to the octave. If we are to be free to modulate also into the keys of C sharp, G sharp, D sharp, F sharp, B flat, E flat, A flat, D flat, we shall require still more. Now, with the human voice, string instruments, and many wind instruments there is no limit set to the number of notes that can be used. With the piano and the organ it is a very different matter. Each note requires a separate lever or key, and the mechanical difficulties in design and execution become very great. In these circumstances it is not surprising that attempts have been made to compromise by using one note to do service for two. The comma is a very small interval, and might surely be ignored. In this way the number of notes required to the octave would be reduced from eighteen to twelve for the selection of keys indicated in the Table. The process of compromise by which true intonation is sacrificed to the exigencies of the mechanics of keyed instruments in order that freedom to modulate may be secured is called temperament. We shall now go on to consider in detail two of the methods by which it has been attempted.

Mean-Tone Temperament.—Only two methods of temperament need be considered—mean-tone temperament for its historical interest, and equal temperament because of its now universal use. The mean-tone system was foreshadowed by Schlick.[1] His suggestion was that the fifths F–c, c–g, g–d', d'–a', should be tuned as flat as the ear can endure, so that a' may make a decent third with f' two octaves above F. The system was given a more precise form by Zarlino in 1562, and later by Francis Salinas of Burgos, who, blind from infancy, was Professor of Music in the University of Salamanca. In his *De Musica libri septem*, published in 1577, he discussed various ways in which the distinction between the major tone and the minor tone could be abolished. As finally adopted the method was so to flatten each of the four fifths from F to a' as given above that the interval f' to a' would be an exact major third. Thus the tempered fifth must be one-quarter of two octaves plus a major third. The major third is approximately 386 or very exactly 386·314. Thus the tempered fifth = $\frac{1}{4}$(2,400 + 386·314) = 696·578. This fifth differs very little indeed from the true fifth, the difference being approximately 5·4 cents or about $\frac{1}{34}$ of a tone. Developing the scale in exactly the same way as Pythagoras, in terms of fifths and octaves, but this time using the tempered fifth, we proceed upwards and downwards by fifths, and bringing all the notes so obtained into the same octave, we have the following scale of twelve notes:

c	c♯	d	e♭	e	f	f♯	g	g♯	a	b♭	b	c'
0	76	193·2	310·3	386·3	503·4	579·5	696·6	772·6	889·7	1,006·8	1,082·9	1,200

Taking the scale of c major from this selection, we see that the seconds in the scale are 193·2, 193·1, 117·1, 193·2, 193·1, 193·2, 117·1. Thus the distinction between the major tone 204 and the

---

[1] *Spiegel der Orgelmacher und Organisten,* 1511.

minor tone 182 is abolished, and a mean tone approximately 193 substituted. The diatonic semitone 112 becomes 117, and the chromatic semitone, which is 92 on the true scale, becomes 76. Comparing the intervals from the tonic on the two scales, we have:

Interval.	Second.	Third.	Fourth.	Fifth.	Sixth.	Seventh.	Octave.
True scale	204	386	498	702	884	1,088	1,200
Mean tone	193	386	503	697	890	1,083	1,200
Error of mean tone	−11	0	+5	−5	+6	−5	0

Thus the largest error is in the second, where it does not matter much, the thirds are true, and the other intervals are only out by about 6 cents or $\frac{1}{34}$ of a tone. Further, this is true of the major scales of C, G, D, A, F, and B♭. If we use the scale E major we must use e♭ to do duty for d♯, and this involves us in an error of 42 cents, or about half a semitone. Similarly E♭ involves us in the use of g♯ for a♭, with a similar error. The false fifth g♯ to e♭, and the false thirds b to e♭, f♯ to b♭, c♯ to f, and g♯ to c are the 'wolves' which haunt the more remote keys and destroy the harmoniousness of the chords by their howling. They can only be exorcised by introducing one further new note for each key made available. So long, then, as we confine ourselves to a few selected keys, the mean-tone scale is admirable. Each of the twelve notes of the scale may be taken as the lower note of an interval, so that there are twelve intervals of each kind which can be reproduced by octaves. Calculation shows that we have nine minor thirds which are 6 cents too small, with their inversions, nine major sixths, which are 6 cents too large; we have eight major thirds and eight minor sixths which are true; and we have eleven fourths 6 cents too large and eleven fifths 6 cents too small. The remaining intervals are out by anything from 36 to 47 cents and therefore unusable.

Such was the system of temperament which came into use in Europe in the seventeenth century and was universal about A.D. 1700. Our modern musical notation, with its distinction between sharps and flats, is based on it. It enables much more harmonious music to be played in six major keys and three minor keys than our present system of temperament allows. It is the

system for which Handel wrote, and for which all the ecclesiastical compositions of Bach were written. It is generally
thought that Bach urged the change from mean-tone temperament to the equal-temperament system which has now completely
superseded it. It is pretty certain, however, that Bach's organ
was tuned in mean tone, from his selection of keys for organ
compositions, if from nothing else. The organ reveals inaccuracies of tuning in a way few instruments do. As for Bach's
48 Preludes and Fugues for the Well-tempered Clavier, these
were published separately in two parts. The first part did bear
in its title the reference to the well-tempered clavier, but in the
second part there is no such reference in the title in the manu-
script, nor in the published edition which appeared in 1799,
fifty years after Bach's death. Bach played little on the pianoforte, and, if the judgement of critics is to be trusted, the clavichord when played loudly so altered the tension, and therefore the
pitch, of the strings as to introduce errors of anything up to half
a semitone. If this were so, the difference between true and
tempered intonation would be completely obliterated. It may
very well be, however, that Bach was in favour of greater freedom
to modulate, and, being familiar with the wolves of the more
remote keys in mean-tone temperament, favoured the introduction of equal temperament without being in a position really to
compare mean tone in its best keys with equal temperament.
He did not convert the contemporary organ-builders, nor did his
son, Emanuel Bach, who advocated the change, make rapid
headway. Not a single organ in the British Exhibition of 1851
was tuned on the equal-temperament system. The first English
organs to be so tuned seem to have gone out from the makers
about 1854, although the Broadwood pianos were tuned on the
new system in 1842.

Equal Temperament.—The system which succeeded mean-tone
temperament was one in which the octave was divided into
twelve equal divisions, each of which was a tempered semitone.
Two semitones were together equal to one tone. All tones were
equally large. The distinction between, e.g., $a\sharp$ and $b\flat$ was
abolished. This is equivalent to making twelve fifths equal to
seven octaves. Thus, ignoring the octaves in which they lie,
we have:

$$f-c-g-d-a-e-b-f\sharp-c\sharp-g\sharp-d\sharp-a\sharp-e\sharp$$

and we arrange to have $e\sharp$ in unison with f. Now, seven octaves

= 8,400 cents, so that the tempered fifth is $\frac{8400}{12} = 700$. Using the same principle of ascending by fifths and bringing the notes into the same octave, we get:

c	c♯ d♭	d	d♯ e♭	e	f	f♯ g♭	g	g♯ a♭	a	a♯ b♭	b	c
0	100	200	300	400	500	600	700	800	900	1,000	1,100	1,200

Here again we have twelve notes to the octave. But this time we have obliterated all differences between the keys. All keys are equally good. We have achieved complete freedom of modulation: what is the price we have paid? Comparing the intervals of the major scale with the true intervals, we have:

Interval.	Second.	Third.	Fourth.	Fifth.	Sixth.	Seventh.	Octave.
Tempered .	200	400	500	700	900	1,100	1,200
True . .	204	386	498	702	884	1,088	1,200
Error of tempered interval . .	−4	+14	+2	−2	+16	+12	0

We see that in this scale there are no true intervals at all, except the octave. The fourth is two cents too large, the fifth two cents too small; but the third is 14 cents too large, and the sixth 16 cents too large. The possible concords within the compass of an octave are nine minor thirds all 16 cents too small, eight major thirds all 14 cents too large, seven perfect fourths all 2 cents too large, five perfect fifths all 2 cents too small, four minor sixths all 14 cents too small, and three major sixths all 16 cents too large.

There is no doubt whatever that these errors represent a real sacrifice of smoothness and harmoniousness in the tempered scale. No one who has first listened to the common chord on the tonic in true intonation and then in tempered intonation can ever forget the experience. And it is the mechanism of the keyboard which imposes the limitation. Violinists and singers are free, when unaccompanied, to use perfectly true intervals. Whether they do in point of fact use them is doubtful. Helmholtz (English Edition, p. 325) quotes Delezenne in favour of the view that violinists do, and adds an account of experiments he made with Joachim which confirmed that view. On the other hand, a much more detailed study of the performance of violinists was made by Paul Greene.[1] Six professional violinists took part, and the major findings were: (a) marked deviations from the tempered intervals, (b) con-

[1] *Journ. Acous. Soc. Amer.*, Vol. 9, p. 43 (1937).

siderable variation in the extent of the deviations, (c) considerable agreement in the direction of the deviation, (d) a mean value for each interval which was much nearer to that of the Pythagorean scale than to the natural scale. The last result is the most surprising. Thus in the case of the minor third we have:

Pythagorean interval . . .	294
Tempered interval	300
Natural interval . . .	316
Average interval played . . .	296

and in the case of the major third:

Natural interval	386
Tempered interval	400
Pythagorean interval . . .	408
Average interval played . . .	406

In the case of singers, Helmholtz (p. 427) speaks with equal certainty. He was greatly impressed by the use of the Tonic Sol-fa method of instruction in England, and convinced that choirs trained by this method sang true intervals when unaccompanied. He had a harmonium tuned in true intervals, on which he accompanied singers at the commencement of a melody, ' and then paused while the singer took the third or sixth of the key. After he had struck it, I touched on the instrument the natural, or the Pythagorean or the tempered interval. The first was always in unison with the singer, the others gave shrill beats.' He concludes (p. 448) that: (a) the intervals which are called ' natural ' are really natural for uncorrupted ears, (b) that the deviations of tempered intonation are really perceptible and unpleasant to uncorrupted ears, and (c) that notwithstanding the delicate distinctions in particular intervals, correct singing by natural intervals is much easier than singing in tempered intonation.

Whatever may be said on this discrepancy of evidence, we must at least bear in mind that it refers to intervals taken in unaccompanied single-part music, and in this case the interval is certainly taken with less precision than when produced as a concord. It is just impossible to believe that anyone listening to the concord of the major third while the tuning is varied through the range from the Pythagorean to the natural interval could fail to notice the increasing roughness as the interval was made to deviate from its natural value. For harmonic music the sacrifice involved in the adoption of equal temperament is a real one, and ought to prompt us to inquire whether it would not really be better to increase the number of keys on the keyboard of the pianoforte or the manual of the organ or, alternatively, to be satisfied with some restriction of the freedom to modulate.

Experiments in the design of more elaborate keyboards have been made—notably by T. Perronet Thompson [1] and by R. H. M. Bosanquet [2] and others. The two specially mentioned based their keyboards on the division of the octave into fifty-three equal parts, called commas. This division was originally suggested by Nicholas Mercator in the seventeenth century. Each comma had thus a value of 22·64 cents. It is possible with very small errors to represent all the more familiar intervals by a whole number of commas:

Interval	Number of commas assigned.	Interval in cents.	True interval in cents.	Difference.
Minor Third . . .	14	316·98	315·64	1·34
Major Third . . .	17	384·91	386·31	−1·40
Fourth	22	498·11	498·04	0·07
Fifth	31	701·89	701·96	−0·07
Minor Sixth . . .	36	815·09	813·68	1·41
Major Sixth . . .	39	883·02	884·36	−1·34

Helmholtz heard music on Thompson's organ on the occasion of a visit to this country, and was very greatly impressed. [3]

Dufton has called attention to the subject again and has designed a 'Harmonodeik', which is shown in Fig. 11.1. [4] The device comprises two concentric dials, the inner one being divided into fifty-three equal parts. In the figure the relative position of the the dials makes B the keynote, and scale of B major may be read opposite the lozenge-shaped indices as B, C♯, D♯, E, F♯, G♯, A♯, B. The scale of B minor may be read opposite the spotted indices, as B, C♯, D, E, F♯, G, A, B. Dufton expresses the view that ' it may well be that the recent advances

[1] *Just Intonation* (1866).

[2] *Musical Intervals and Temperament*, Macmillan and Company (1876).

[3] For further reading on this fascinating subject the reader is referred to J. M. Barbour's very thorough treatise *Tuning and Temperament* published by the Michigan State College Press in 1951.

[4] *Phil. Mag.*, Vol. 32, p. 260 (1941).

in physics will enable us to safeguard the principles upon which musical art is founded and to avert the threat that the mechanism of our present instruments and attention to their convenience will lord it over the natural requirements of the ear.'

Tuning of Keyed Instruments.—It is one thing to agree on a tempered scale; it is quite another thing to realize this scale on the keyboard of a pianoforte or organ. To tune two notes to form an exact octave by ear is fairly easy, and to tune a perfect fifth, as the violinist does, is comparatively simple. The correct

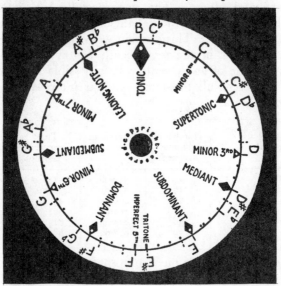

FIG. 11.1.—Harmonodeik

interval is marked by a notable minimum roughness. But the tuning of an imperfect fifth—imperfect by just the right amount—is very much harder, and quite impossible to achieve by an æsthetic judgment of the ear alone. Of course the tuning might be accomplished by carrying round a set of twelve tuning-forks, accurately tuned to the twelve notes of one of the octaves of the tempered scale. This series of notes on the pianoforte would then be tuned to the forks, and the remaining notes of the pianoforte keyboard tuned by true octaves. In practice this is never done. One note is tuned in unison with a standard fork; the remaining notes within an octave round this are tuned by tempered fifths and fourths, and the rest of the tuning is done by true octaves.

PLATE XIII

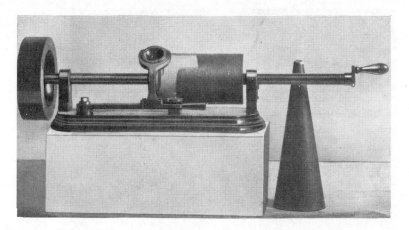

FIG. 12.1.—Photograph of early type of Edison machine for recording and reproducing

FIG. 12.2.—Berliner's Gramophone, 1894

PLATE XIV

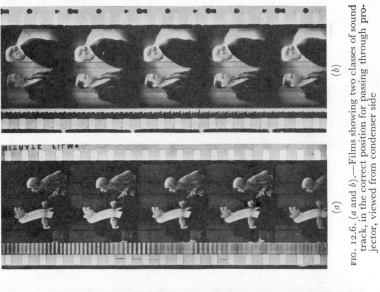

FIG. 12.6. (a and b).—Films showing two classes of sound track, in the correct position for passing through projector, viewed from condenser side

FIG. 12.3.—Marconi-Stille Recording-reproducing equipment. Type MSR3.

Interest centres chiefly in the tuning of the pianoforte. In the case of the organ and harmonium the same procedure is followed, but the sustained tone of these instruments simplifies the task and increases the accuracy attainable. In the case of the pianoforte, the artiste formerly had to do his own tuning, and the performance had frequently to be suspended for this purpose. Now the steel frame and other improvements have made the tuning more permanent, and the technique has been relegated to specialists, whose methods are often completely unknown to the pianist.

There are four steps in the process:

(*a*) One of the wires of one note must be tuned to the standard fork;

(*b*) One wire of each of the notes in the octave round the first one must be tuned by proceeding from the first by a series of tempered intervals;

(*c*) One wire of each of the remaining notes must be tuned from those by a series of true octaves;

(*d*) The remaining wires for each note must be tuned by unison.

Rubber keys or corrugated flannel are used to prevent all except the selected wires from vibrating. The process (*b*) is known as ' laying the bearings '. Methods of tuning are discussed in detail in W. B. White's *Modern Piano Tuning*, and there is an excellent article on the subject by G. F. H. Harker.[1]

The first step is to tune a' to a frequency 440. This is achieved by eliminating the beats between the wire and the fork. The tuner can then proceed from a' to d'. The tempered frequency of d' has been calculated, and found to be 293·6648. The third partial of this is 880·9944, while the second partial of 440 is 880·0000. The beat-frequency is 0·9944 per second, or 10 beats in 10·1 seconds. This beating between the partials is fairly distinct and easily counted, especially with the front of the piano removed or, in the case of the grand piano, the top lifted. The result is most accurate when the rate of beating is moderately slow and the prolonging of the tone sufficiently great to enable a reasonable number to be counted. These conditions obtain in the neighbourhood of a'. The beat-frequency is about one

[1] *Journ. Acous. Soc. Amer.*, Vol. 243, p. 8 (1937).

198 THE PHYSICS OF MUSIC

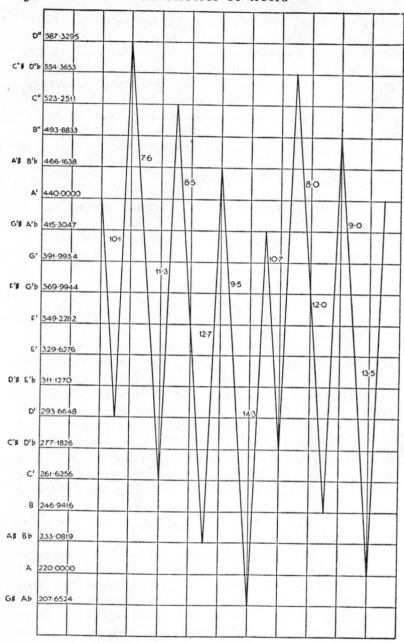

FIG. 11.2.—Graph. Laying the bearings for the tuning of a pianoforte by descending fifths and ascending octaves

per second, and the tones are sustained sufficiently for beats to be counted for ten to twenty seconds. In the bass the tones are sustained longer, but the rate of beating is too slow. In the treble the beat-frequency is higher, but the damping of the vibration of the string occurs very rapidly, and the partial tones decay more rapidly than the fundamental. The beats are, of course, most distinct when the digitals are struck so that the two partials whose beating is to be timed have as nearly as possible the same intensity. The result of the process is shown in Fig. 11.2.

Thus we start and finish with a', and in the process we move by twelve descending fifths and seven ascending octaves. There is obviously something to be said for starting at the two ends and meeting in the middle, instead of letting the errors accumulate until we return to the note we started from. Alternatively we could, of course, have used descending octaves and ascending fifths.

Although fifths are traditionally used for tuning, fourths are equally good. It is easier to tune a true fifth than a true fourth, but it is no easier to tune a tempered fifth than a tempered fourth. The beat-frequencies are equally suitable. In the perfect fourth the ratio of the two frequencies is $4/3$. Thus the fourth partial of the lower note coincides with the third partial of the upper. Taking a descending fifth from a', we have e'. The fourth partial of e' is $4 \times 329 \cdot 6276 = 1318 \cdot 5104$, while the third partial of a' is $3 \times 440 \cdot 00 = 1320 \cdot 0000$. The beat-frequency, being the difference between these two numbers, is $1 \cdot 4896$ beats per second, or $14 \cdot 9$ beats in ten seconds. This compares with $9 \cdot 9$ in ten seconds for the corresponding fifth. In the case of the fourths the procedure is $a'–e'$, $e'–b$, $b–b'$, $b'–f'\sharp$, $f'\sharp–c'\sharp$, $c'\sharp–c''\sharp$, $c''\sharp–g'\sharp$, $g'\sharp–d'\sharp$, $d'\sharp–a\sharp$, $a\sharp–a'\sharp$, $a'\sharp–f'$, $f'–c'$, $c'–c''$, $c''–g'$, $g'–d'$, $d'–d''$, $d''–a'$. In this process there are seventeen steps (twelve descending fourths and five ascending octaves), as against nineteen in the other. The highest number of beats per ten seconds is $19 \cdot 9$ in the interval $d''–a'$, and the smallest number is $9 \cdot 4$ in the interval $c'\sharp–g\sharp$. Here again we can execute the same plan by using ascending fourths and descending octaves, and in either method we may start from the two ends and meet in the middle.

A more elaborate method of laying the bearings is given by W. B. White,[1] in which the steps are fifths and fourths and there is a continuous check by thirds and sixths.

---

[1] *Journ. Acous. Soc. Amer.*, Vol. 47 (1937–8).

Harker points out that the timing of the beats may be carried out in three different ways: (a) by adjusting some form of metronome to coincide with the beats. The metronome would have to be carefully compared with a clock, and some modification of existing types would greatly increase their usefulness for

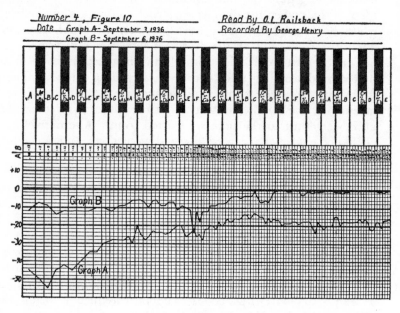

FIG. 11.3.—Chart showing results of tuning pianoforte. tuning. Graph A shows condition one year

this purpose. (b) By counting the beats in a given time. It is difficult to do this accurately, as it is impossible to estimate the fraction of a beat with any precision. (c) To time a given number of beats with a stop-watch reading to fifths of a second. The last method is the one likely to give the greatest accuracy, but it is rarely used. Indeed, Harker doubts if piano-tuners ever count beats! He thinks they tune to the true interval and then contract it or expand it by a 'wave' (presumably one beat per second), or tune their intervals 'slightly' or a 'trifle' or a 'shade' flat or sharp. Certainly without the counting of beats no real accuracy can be relied upon. On the other hand, a craftsman who spends his life performing a limited series of operations may attain an unexpected degree of precision, and this may be true of the piano-tuner. Grutzmacher and Lottermoser have tested the work of the tuner, and found errors up to one-tenth of a semitone (about 10 cents). An interesting pair of graphs was published by Railsback[1] and is shown in Fig. 11.3. They show (Graph A) the condition of a piano about one year

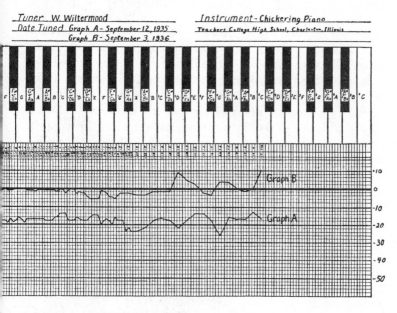

Graph B shows condition of the pianoforte three days after
after tuning. The vertical scale is in cents

after tuning. The pitch in the extreme bass has fallen by about a quarter of a tone. The middle of the keyboard is not so bad, but is round about one-tenth of a tone flat. In the treble the average error is about the same, but the variation of groups of neighbouring notes is much greater. The second graph (B) shows the condition of the same piano three days after re-tuning. The average error in the bass is now only one twentieth of a tone. Over the middle register the tuning is very nearly exact. In the treble the notes are, on the whole, sharp, and show the same kind of variation as before tuning.

[1] *Journ. Acous. Soc. Amer.*, Vol. 9, p. 37 (1937).

CHAPTER 12

RECORDING AND REPRODUCTION OF SOUND

Recording and Reproduction of Music.—Recent scientific developments have made it possible to record music, store the record, and
reproduce the music at will. It may be argued that reproduced
music is never the same as music heard when first produced, but
the reproduction of music does open up possibilities of very great
importance. From the point of view of the ordinary listener it
makes available the kind of music he wants at the time he wants it.
Broadcasting has gone some way to meet this demand. The wide
variety of stations and the wide selection of programmes which can
be picked up on a moderately good radio set give considerable
scope, but it still often remains true that the mood of the moment
cannot be satisfied. From the point of view of the student there
is an added advantage. A particular passage can be repeated
indefinitely until it has been thoroughly analysed; comment and
criticism can be exchanged, and the process can be extended until
a complete work has been mastered. The reproduction of inci-
dental music with films is also an important development which
these scientific developments have made possible.

Of course, it is undoubtedly true that usually the quality of
reproduced music leaves much to be desired, but in reply to
this criticism two things can be fairly said: (1) the quality is all
the time steadily improving, and in the case of the best gramo-
phones has now reached a very satisfactory standard, and (2) the
quality would be improved still further if the public were more
critical. In some cinemas, the sound-reproduction unit is
miserably inefficient compared with the best technical standards,
and could be improved out of all recognition. So long as the
public seems satisfied, however, no more money will be expended
and no more trouble taken to make improvements on the sound
reproduction which would not be reflected in box-office receipts.

Recording on Wax.—Probably the earliest successful attempt to
obtain some kind of record of the air vibrations responsible for a
sound was made by Leon Scott in 1857. His apparatus was
subsequently patented as the Phonautograph, and improved by

Koenig. The sound to be recorded was collected by a large parabolic horn, the narrow end of which was closed by a membrane. To this membrane a bristle was attached by a lever in such a way that when the membrane was at rest it traced a uniform line on a revolving cylinder covered with lamp-black. When the horn was directed to a source of sound, the bristle vibrated parallel to the axis of the cylinder and made a wavy trace. It was originally designed and used to measure the frequency of a source of sound. It could not be used to obtain a trace from which the sound could be reproduced.

The next important step was taken by Edison, who in 1877 adapted the phonautograph so as to obtain from it a record capable of reproduction. The apparatus is shown in Fig. 12.1, Plate XIII. The membrane which closes the narrow end of the funnel, in this case conical, carries a very narrow chisel which indents the material placed on the cylinder to obtain the record. The motion of the chisel is such that when the membrane is exposed to sound waves the chisel cuts a furrow of varying depth instead of moving from side to side across the surface of the cylinder as in the case of the phonautograph. A screw ensures that the cylinder not only rotates as the handle is turned, but moves along parallel to its axis. Thus the chisel cuts a helical line on the record material which is of variable depth if the membrane is set in motion by sound waves. At first the record was made on tinfoil, and was reproduced by replacing the chisel at the beginning of the groove and making it retrace its path by turning the handle. The up-and-down motion of the point in the groove is communicated to the membrane, which is set in vibration and repeats the motion by which the groove was originally traced. The vibration of the membrane sets up in the air in the horn the pressure changes which originally produced the vibration—i.e, reproduces the original sound. Bell recorded on wax cylinders using a sharp stylus and stiff membrane, and reproduced by using a blunt stylus and limp membrane. He also duplicated records by an electrotype process similar to that still in use, the wax record being cast in a mould. Some of his equipment and many cylinders, including paper-based ones, may be found in the Alexander Graham Bell Museum at Baddeck, Nova Scotia. In 1878 Edison produced the phonograph, the first practical 'talking machine', and this was introduced to the Royal Society of Arts on May 8 in that year. The air vibrations were conducted by rubber tubes to the two ears of each listener.

In 1887 the Berliner gramophone was produced (Fig. 12.2, Plate XIII). The record was made on a disc, and the vibrations were ' lateral '—i.e., side to side across the groove—instead of ' hill and dale '—i.e., up and down in the groove. The disc was of metal and was covered with wax. The recording stylus removed the wax and the trace was bitten in with acid. The gramophone was hand-driven and was equipped with a very small horn. In 1898 there appeared a new and improved model. It was driven by clockwork and had a rather larger horn ending in a flare. This model is immortalized in the famous picture by Francis Barraud which suggested to the Gramophone Company the name of ' His Master's Voice ', and led to the model being known as the Dog Model.

Meantime better materials were being used for records and the process of recording improved in many ways. In 1924 a very important advance was made. Until then the recording tool was operated with power derived directly from the source of sound. Thus only fairly intense sounds could be recorded. The sound had to be collected by a large horn, and the soloist had to be situated right in the mouth of the horn.[1] An orchestra or band had to be crowded round the end of the horn in circumstances of the greatest discomfort. Also the recording mechanism was too clumsy to be moved about. But now the process of electrical recording was introduced. It was made possible by the discovery of the thermionic valve in association with a microphone. No attempt will be made here to explain the action of these devices. The essential function of the microphone is to transform the changes in pressure of the air in front of its diaphragm into identical changes in an electric current, so that every variation in air pressure is linked with a corresponding variation in an electric current. The function of the valve is to amplify this current without distortion—i.e., increase its magnitude without altering it in any other way. This amplified current can then be applied to the tracing tool, and the power necessary to move the tool is drawn, not from the source of sound, but from the amplifying system. As very great amplification without appreciable distortion is available, very feeble sounds can be recorded, the

[1] An amusing account of this procedure may be found in Gerald Moore's memoirs *Am I too Loud?*

microphone may be some distance from the performers, and records may be made anywhere, as the microphone may be in one place and the recording mechanism in another.   In this way it is possible to get a motion of the recording stylus which is a very accurate reproduction of the motion of the air.   The departure from accuracy resolves itself finally, of course, into a difference in quality.   Either there is a change in the relative intensities of the partial tones, or new partials are introduced, or the distortion is a mixture of both.

Magnetic Recording.—Instead of recording and reproducing sounds by means of a template on a disc or cylinder which a stylus follows, one can also record sounds as a pattern or arrangement of magnetic domains in a suitable medium.   Recording on a magnetic medium was invented by Oberlin Smith who took out a patent in 1888 in the U.S.A. Poulsen used steel tape in 1900 and exhibited his Telegraphone at the Paris Exhibition in that year. In 1924 research on the process was begun by Dr. Stille, and later the Blattnerphone was produced, based largely on Stille's patents, although certain improvements were made by the BBC engineers in collaboration with the British Blattnerphone Company.   In 1933 the system was further improved by the Marconi Wireless Telegraph Company, and one of the Marconi-Stille recorders is now preserved in working order in the Science Museum, South Kensington.   This machine is of great interest to all who have anything to do with sound recording and it is well worth while to make an appointment to see it working (Fig. 12.3, Plate XIII). A Blattnerphone was revived for the 50th Anniversary celebrations of the BBC in 1972 with commendably good results for its age of about 40 years.

The record is made by passing a tape of steel or some magnetic material past a strong electromagnet which magnetizes it to saturation.   This magnetization is then modified by driving the tape past a second electromagnet arranged to reduce the magnetization of the tape, and excited by the variable current from the receiving microphone.   In this way the magnetization of the tape is made to vary from point to point, the variations following the variations of the current from the microphone, and therefore variations of pressure in the sound wave.   If therefore the tape is now moved past a coil it will stimulate a variable current in the coil, the variations of current following the variations in the strength of magnetization of the tape.   The variable current is then fed to a loudspeaker and the reproduction is complete.

The method has some important advantages. (1) The recording is ready for immediate playing, and this makes it very suitable for use in broadcasting. An event can be recorded and several repetitions given in the various news broadcasts of the day. It can also be included in a summary later in the week, or kept for future use. (2) If the record is not required for future use, it can be used for a new recording. When the tape is driven past the magnetizing head it is ' wiped ' clean, and is ready for use again. Thus the same length of tape can be used many times. (3) It is very little affected by mechanical vibration and shock.

The response of the system described above was fairly uniform for frequencies from about 70 Hz to about 4,000 Hz, and so the quality of reproduction was quite good. Continued playings have very little adverse effect on the volume obtained from the record, the effect of 200,000 playings being a drop of less than 5 phons.

In the Marconi-Stille recorder, the tape was first biased to saturation by an electro-magnet carrying D.C., and the signal was applied to reduce the magnetization. This method has been found to be inferior to the method of A.C. biasing both in harmonic distortion and signal to noise ratio. The signal to noise ratio is the ratio of the peak signal level at some specified distortion to the total noise level integrated over the usable frequency bandwidth of the system. For example, a tape recorder may have a signal to noise ratio of 300 (or 50 dB) when peak signal level is defined as that corresponding to 2 per cent. harmonic distortion and the noise is integrated over a bandwidth of 50 Hz to 12,000 Hz.

It is not appropriate to attempt a detailed discussion of the physics of the A.C. biasing method here as it is a subject of some difficulty and is better treated in a more specialized text. Briefly, the signal to be recorded is added to a high frequency bias signal and both are applied to the recording head of the tape recorder. The magnetic medium—the tape—is thus subjected to a high frequency magnetic field whose amplitude is determined by the bias level used (which is adjusted to suit the characteristics of the medium) and whose mean level is determined by the signal to be recorded. The interaction of the high frequency field and the recording medium is complex, but as a result of the anhysteretic properties of the latter it is found that a very close approximation of the original signal is impressed on the tape in the form of variations of magnetization.

The actual tape used has also been changed—no longer do tape recorders have huge spools with a mile or more of steel tape on each for twenty minutes' recording time as may be found on the Marconi–Stille machine. Instead a plastic-based tape is used coated with a thin layer of various magnetic oxides; by changing the proportions of different oxides used, the tape may be given a range of properties according to its intended use. For example, it may be designed to have low ' print-through '. Print-through is one of the major disadvantages of the method of magnetic tape recording and arises in the following way: after a signal has been recorded, the tape on which it has been placed is wound on to a spool. Now if the tape is fairly thin, the variations of magnetization that compose the signal on one layer of tape may be able to impress themselves on the adjacent layers so that when the tape is replayed loud signals are clearly preceded and followed by ' echos ' of themselves. This phenomenon gives rise to alternative names for print-through—pre-echo or post-echo. Of the two, pre-echo is the most disturbing to the musician, apparently because it does not have a counterpart in nature, and many otherwise excellent recordings of music have been ruined by its presence. Unfortunately a tape that has low print-through is inferior in other respects to a tape of normal design, and commercial tapes represent a range of compromises which are designed to suit the different users.

One special feature of magnetic tape recording is the ease with which the recording may be ' edited ' afterwards. A poor section in a recording may be simply cut out and replaced with another. Indeed, this editing is so easy that one no longer knows whether a recording of a performance corresponds to just one performance of the music or is made up of twenty or more different performances, combining the best parts of each. Out of this editing has grown a new art form—' electronic music '. Electronic music exists in various forms, but all depend ultimately on the fact that a sound recorded on tape may thenceforth be regarded as a ' sound-object '—a concept due to Pierre Schaeffer [1] —and may be manipulated and transmuted at the will of the composer.

Disc Records.—A performance of music is thus recorded on a

[1] Pierre Schaeffer—*Traité des Objets Musicaux.* Edition Leseuil, Paris 1966.

tape which may or may not have been edited by the recording
engineers; as discs still comprise the major outlet for the commercial sale of recorded performances, the tape record must be
converted to one on disc. A disc of soft material with a very
smooth surface is placed in position on a turntable under the
recording point. As the disc rotates on the turntable the recording point is moved steadily towards the axis of the table, so that
the stylus records a smooth spiral line from the outside edge of the
disc towards the centre. If now the electric current from the
tape playback machine is applied to the recording mechanism,
the stylus is made to vibrate to and fro, along a radius of the
disc, and so to swing from side to side across its spiral track,
cutting a wavy line instead of a smooth spiral. The next stage of
the process is to pass from a very soft material for recording to a
harder substance for the final record. The soft record could be
played, but it would deteriorate with great rapidity. The surface
of the selected record is first rendered electrically conducting by
brushing with graphite or otherwise. It is then immersed in a
bath of copper sulphate solution, through which an electric
current is passed from a copper electrode to the surface of the disc.
In this way the disc becomes electroplated with copper, the copper
being deposited at first as a very thin film, gradually thickening
into a shell which can be stripped from the record. This shell—
the master shell—is a negative which fits exactly on the record,
the groove of the record appearing on the shell as a continuous
wavy spiral ridge. Obviously this shell could be used for stamping records which would be ' positive '—i.e., they would have a
groove identical in all respects with the groove on the original
record. This would soon wear the shell, however, and, in order
to avoid this, the master shell is treated in the same way as the
original record—put in a bath of copper sulphate solution and
electroplated with copper. The process is arranged so that the
copper shell so produced does not adhere to the master shell, but
can be stripped from it, giving the mother shell identical in all
respects with the original record. This in turn is electroplated,
and gives the matrix shell, a negative again, which can be used for
stamping. This shell, after nickel plating, is prepared and
mounted on a heavy copper disc, and the central hole is accurately bored. Matrices, having been prepared for each side
of the record, are fixed in position in a hydraulic press on plates
which can be alternately heated and cooled. Record material
(a mixture of shellac, copal resin, carbon black, and slate powder

for 78 r.p.m. records) is placed on a hot table and softened to the required plasticity.   A lump is then put on the lower face of the press, the upper face is brought down on top, and hydraulic pressure is applied.   Then after the dies have been steam-heated, cold water is passed through till the record is sufficiently cooled. With ' long-play ' records a smoother vinyl plastic is used which is softer than shellac but has a smoother, less noisy, surface, and is eminently suited to the very light-weight pickups that were generally introduced in the early 1950's.   After being pressed, the record is removed, polished, tested, and is available for use.

Playing Time.—The main disadvantage of the early shellac records which rotated at about 78 r.p.m. was the shortness of the playing time.   Developments in pick-ups of light weight and high compliance made it possible in 1950 to introduce ' long-play ' records which have smaller grooves than the older 78 r.p.m. records, and rotate more slowly, at $33\frac{1}{3}$ r.p.m.   As mentioned above, these records were made of a new material with a less noisy surface so that the maximum signal to be recorded could be reduced without spoiling the signal-to-noise ratio.   In fact, the signal-to-noise ratio of a vinyl record is appreciably better than that of a shellac record even with the lower recording level. Since the maximum amplitude of swing of the needle has been reduced by using a lower recording level, the grooves can be packed closer together.   These three factors: smaller grooves, slower speed, and grooves packed closer together combine to enable one to record up to half an hour continuously on a 12″ diameter record as opposed to about five minutes on a 12″ diameter shellac record rotating at 78 r.p.m.

Multichannel Recording.—With magnetic tape it is very easy to record and play back more than one channel at a time.   In many recording studios it is common practice to record each section of a band or orchestra on a separate channel and to perform the 'mixing' of these channels into one or more outputs in the relative calm and quiet of the recording room after the musicians have left the studio.

Multichannel recording is carried out on tape by using more than one recording head on the tape; instead of magnetizing the whole width of the tape, the recording head may magnetize only slightly less than half the tape width, another head magnetizing slightly less than the other half.   Thus we would have two-channel recording with a little gap left in between to reduce interaction between the channels.   If the two heads are arranged so that the gaps left in their magnetic cores are colinear, the recording is known as ' in-line '; if the two heads are spaced along the tape, the recording is called ' staggered '. In-line recording is more common and staggered is only used when there are great difficulties with interaction between channels.

More than two channels may easily be arranged for; though it is unusual to find professional equipment with more than two channels on $\frac{1}{4}''$ wide tape, domestic equipment frequently has four so that greater use may be made of the tape. Wider tape may also be employed if more channels are required. For example, eight top quality channels may be accommodated on $1''$ wide tape.

' *Stereophonic* ' *Recording*.—This is a special type of multichannel recording where two or more channels are employed to give the listener more than one auditory ' view ' of the performance. For example, with two channels, one channel may convey the music appropriate to the left-hand side of the ensemble and the other that appropriate to the right-hand side. The effect on the listener who listens through two loudspeakers, or with headphones connected so that each ear hears only one channel, is, in favourable circumstances, as if the ensemble can be perceived in perspective and the sound source is spread out over an angle of view corresponding to that existing in a live performance.

With growing commercial interest in stereophonic reproduction, much research has been devoted to finding a satisfactory way of putting two channels on a disc. A simple way that immediately suggests itself is to have two bands on the disc, one for each channel of the music. These bands are replayed with a special double pick-up, the styli of which are accurately spaced the exact distance apart of the two bands and are maintained in exactly the same angular relation to the grooves as those on the cutter which recorded the disc. If this is not done the two channels will get out of step and the stereophonic illusion will be destroyed. The disadvantages of this method are obvious: firstly, the playing time of any sized disc will be halved; secondly, the system is not compatible—the special pick-up can only be used for these special discs; thirdly, the mechanical accuracy required in both the pick-up and its arm is very great.

A solution seems to have been found to these problems by reviving Edison's original method of recording on a cylinder, in which the recording stylus did not move from side to side and cut a wavy groove, but moved up and down cutting a groove of variable depth—hill-and-dale recording.   In this method of two-channel recording the recording stylus moves both up and down and from side to side, enabling the two channels to be recorded in the same groove.   Because it has been found that the distortion in a hill-and-dale recording is approximately three times as much for the same recording level as for a side to side recording, the two motions of the stylus used are not the simple up and down for one channel and side to side for the other.   Instead the directions of cutting are inclined 45°, thus distributing the distortion between the two channels.   This method of recording is, appropriately enough, called ' 45–45 ' recording.

45–45 recording is compatible, the pick-up can be used to play ordinary, single channel, records; the playing time is unaltered; and the pick-up does not have to be specially aligned to the disc to ensure that the two channels replay simultaneously.

Recently interest has grown in the reproduction of music through four channels rather than two, the idea being that the music can be apprehended by the listener ' in the round ' with all the ambience of the concert hall.   Since none of the systems being put forward commercially at present attempt to reproduce the information coming from above the listener in a concert hall (which forms a significant part of the ambience of the hall, as a short test will readily verify) some doubts are expressed by serious listeners as to the motives of supporters of ' quadraphonic ' reproduction as it is known. It is generally agreed that the disc will continue to be the major source of reproduced sound but there are many problems in recording four channels on a disc and several firms are using a ' matrix ' system wherein the four channels are combined into two when recording and on playback the reproducing system attempts to sort out the signals with, so far, not entirely satisfactory results.

Mechanical Reproduction.—Although mechanical reproducers are of mainly historical interest, it is instructive to consider the way the mechanical gramophone worked.

The soundbox is the transformer which converts the swings of the needle point into changes of air pressure in the throat of the horn.   Its essential parts are shown in Fig. 12.4.   It consists of a thin diaphragm, which may be mica or aluminium or a metal alloy, which is set in vibration by a stylus lever connecting the needle-holder to the centre of the diaphragm.   As the aperture of the throat is much less than the area of the diaphragm, the velocity in the throat is relatively high, and the pressure generated much greater than would be the case if the diaphragm had the same diameter as the throat.   This means that the diaphragm works against a higher pressure, more work is done, and more sound energy generated.   In some cases the vibrations are conveyed not to the centre of the diaphragm, but to a series of points on the circumference of a circle on the diaphragm by a multiple lever called a ' spider '.

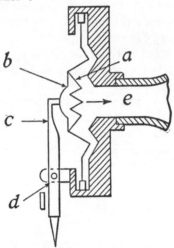

FIG. 12.4.—Sound-box. *a*, corrugated diaphragm; *b*, spider;
c, stylus lever; *d*, needle holder; *e*, throat

It is desirable that the natural frequency of the diaphragm should be high, otherwise when a high frequency vibration is imposed upon it by the stylus lever it breaks up into vibrating parts, with consequent inefficiency.

In the case of a mechanical gramophone the horn is an essential element for effective reproduction.   The vibrations of the diaphragm of the sound-box, if communicated direct to the air, give very poor results.   Only a small fraction of the energy of the diaphragm actually appears as pressure variations in the air. In the early days of the gramophone the function of the horn was not understood.   The theory of its action was only fully developed between 1919 and 1925.   Its function is to increase the resistance against which the diaphragm works, and to transform waves in which the air velocity is high, and which are distributed over a small area (the throat of the horn), into waves in which the air velocity is low and which are distributed over a large area (the open end). The essential features of an ideal horn are: (*a*) small throat, (*b*) large opening, (*c*) suitable form, (*d*) slow taper.

(*a*) The disadvantages of a very small throat are that the frictional resistance to the motion of the air is greatly increased, and that for a given rate of expansion of the horn the smaller the throat, the longer must be the horn if it is to end in an opening of a specified area. The actual area of the throat may be about 600 mm².

(*b*) It can be shown that the less the area of the open end of the

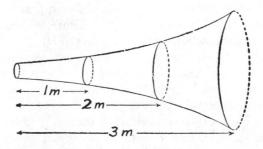

FIG. 12.5.—Exponential horn with rate of expansion 1 metre, *i.e.*, area of cross section doubles for each increment of 1 metre measured along the axis

horn the less bass does it transmit. In fact, a horn may be said to cut down all wavelengths which are more than about one quarter of the circumference of the open end. Now, the ear is sensitive down to frequencies of about 25 Hz. This frequency, however, marks the limit of sensitiveness, and no attempt is made to record it or to reproduce it. We shall impose a sufficiently large demand if we attempt to deal with a frequency of about 112 Hz. The velocity of sound is about 340 metres per second at ordinary temperatures, and this means a wavelength of $340/112 = 3$ m, and therefore a circumference for the open end of 12 m and a diameter of $12/\pi = 4$ m.

(*c*) *Shape of horn.*—The early horns were conical, but it was found that this shape was much less efficient than the so-called ' logarithmic ' or ' exponential ' horn. The flare of this horn is such that the area of cross-section doubles itself at constant intervals measured along the axis of the horn. This gives the outline shown in Fig. 12.5.

(d) *Degree of flare.*—For convenience it is obviously an advantage to have a short horn. If therefore we must have a narrow aperture at the sound-box and a wide opening at the other end, then a horn of reasonable length must flare rapidly from one end to the other. Unfortunately the rate of expansion of the horn is also important, and low frequencies are only radiated by a horn which expands slowly. It follows that for a good bass the horn must be long. Theory shows that to radiate effectively the wavelength of 3 m, the area of cross-section must double itself about every 165 mm. Now, we start with an aperture of 600 mm² at the narrow end, and we finish with an aperture of $\pi \times 4^2$ m² at the open end. The ratio of these two areas is $\pi \times 16 \times 10^6/600$, and it will be found that this requires a horn whose length is about 2·7 m. Thus for nearly perfect reproduction down to frequencies of even 110 Hz, the horn ought to be a straight exponential horn of about 2·7 m length and 4 m diameter at the open end. Fortunately, the ear is very tolerant, and owing to its property of producing difference tones corresponding to the successive pairs of partial tones of a musical note, it provides a bass of its own even when no bass is supplied by the gramophone horn. But for this fact the gramophone industry might well have been strangled at birth. The tiny horns used in the early instruments filtered out all the bass, and if the ear had not supplied this defect, both music and speech would have been wellnigh intolerable.

In order to avoid the inconvenience of very long horns with very wide ends, special forms have been devised—e.g., bent horns, folded horns, and bifurcated horns. All these are compromises, and although they are better than the short horn with small open end, they compare unfavourably with the straight exponential horn of the required length and size.

Electrical Reproduction.—Electrical reproduction has been implied in some of the previous discussions on recording in this chapter, but it is convenient to set out the method of operation here, directly after the section on mechanical reproduction. The variable electric current from the pick-up (or from the tape head) is amplified by a valve or transistor system and delivered to the loudspeaker. The loudspeaker is a reversed microphone— i.e., the microphone converts sound waves into electric current, while the loudspeaker converts a variable electric current into sound waves.

This latter conversion may be made by transmitting the vibrations of the loudspeaker diaphragm direct to the air. This method, as has been pointed out in the case of the acoustic gramophone, is very inefficient, and with a small diaphragm gives very little loudness, even with large amplitudes. The choice then is between a loudspeaker with a comparatively large diaphragm radiating sound directly into the air or with a small diaphragm radiating through a horn, as in the case of the acoustic gramophone. Loudspeakers based on the large diaphragm have been developed mainly for home use, owing to the inconvenient size and considerable cost of large horns. For efficiency the diaphragm ought to be set in a baffle board—a board surrounding the diaphragm—in order to prevent circulation of the air between the front and back. If the air can move from front to back while the diaphragm moves out, and from back to front while the diaphragm moves in, then it can ' circulate ' without sustaining the pressure changes necessary for the transmission of sound waves. In order that this may be avoided, the diameter of a circular baffle board or the side of a square one must not be less than half a wavelength for the lowest frequency to be radiated—i.e., not less than 1·5 m for the limit previously considered.

The commonest loudspeaker using the large diaphragm and no horn is probably the cone diaphragm driven by a moving coil mechanism. A fairly uniform response between frequencies of 100 and 3,500 Hz can be obtained by this type of instrument, and if very carefully designed it may have a range from 80 to 10,000 Hz. Its efficiency is small (something like 1 per cent.), but with the necessary amplification large power can be radiated without appreciable distortion.

The horn loudspeaker is a much more efficient instrument. With a long exponential horn it may be made to give an efficiency of something like 30 per cent. over a considerable range of frequency. The disadvantages of the horn have already been dealt with, as there is no difference in principle between the use of the horn for mechanical and for electrical reproduction.

Recording on Film.—The possibility of obtaining a record of sounds on a photographic film seems first to have suggested itself about 1900. In that year a physicist called Ruhmer designed an arrangement whereby sound-waves could be made to vary the brightness of the light from an arc lamp while a film passed behind a narrow slit illuminated by the lamp. In this way a strip of film was produced which, when developed, showed a varying density depending on the varying brightness of the arc lamp, and therefore on the sound-waves. An electric circuit was then set up, including a loudspeaker and a light-sensitive cell whose electrical resistance varied with the brightness of the light falling on it. If now the film was moved behind a slit illuminated by a steady bright source and the light transmitted by the film was received on the cell, then the variations in the light caused variations in the electrical resistance of the cell, and therefore in the electrical current through the circuit, and these variations, acting on the loudspeaker, became audible as sounds.

In 1906 Eugene Lauste applied to the British Patent Office for a patent for a 'new and improved method of and means for simultaneously recording and reproducing movements and sounds'. The patent was accepted on the 10th August, 1907, but it expired before it was worked. It appeared at a time when the development of the silent film was carrying all before it, and when the unamplified voices were too thin to please an audience in a large hall; Lauste was twenty years ahead of his time, and twenty years is the period of protection for a patent. In 1923 de Forest took out patents for recording on film and reproducing with amplifiers; a demonstration was given at Finsbury Park Cinema, and in 1927 commercial backing was obtained on a large scale. The première of 'The Jazz Singer' in September 1928, and its subsequent run at the Regal Theatre, Marble Arch, marked the beginning of a spectacular development. In a matter of weeks the silent film was completely superseded. The demand for the new sound-film apparatus far exceeded the supply, and skilled operators were unobtainable. Recently one of the patent attorneys of the Bell Telephone Laboratories came across a letter in the press 'complaining that an old man, Eugene Lauste, who had pioneered in the art of talking motion-pictures was destitute while the industry prospered'. He traced the letter and verified its correctness. 'Eugene Lauste then became a member of the technical staff of Bell Telephone Laboratories with two tasks, one to assemble such of his own apparatus as could be located and by replacements to reconstruct the entire system which he had built in the early 1900's, and the other to assist the patent department by his knowledge of the prior art. During the Second World War a complete reproduction of his system for sound pictures was presented to the Smithsonian Institution and accepted. It contains the first light-valve, and also various electro-mechanical devices with which he attempted to get the amplification without which his system, although operative, was commercially impractical.   He was an inventor ahead of his time; but in his latter years his work became known, and he was honoured by his younger colleagues, with, for example, an honorary membership in the Society of Motion Picture Engineers.'[1]

In recording on film, the essential thing is that the light passing through a slit should vary in such a way that its changes correspond to the pressure changes constituting the sound-waves in the air.   There are two ways in which this may be secured.   In the first the whole length of the slit is illuminated, but the brightness of the illumination is varied, either by using the variable current from the microphone to vary the brightness of the source—usually a glow-lamp—or else by using the current to operate a valve placed between the source and the slit.   In either case if a film is passed behind the slit at constant speed and afterwards developed, the density will vary from point to point.   This method is called the 'variable density' method of recording, and the result is shown in Fig. 12.6a, Plate XIV.   An alternative method of recording uses the variable current from the microphone to vary the length of the slit illuminated by a constant source.   In this case, if the film is passed behind the slit and subsequently developed, it will show a transparent edge and an opaque edge with a boundary separating the two which will vary in position from point to point, as shown in Fig. 12.6b.   This method is called 'variable-width' or 'variable-area' recording.

This variable-width or variable-area record can, of course, be imitated by using a strip of opaque film with one edge straight and the other serrated.   This serration we can cut in any way we please.   For instance, we can cut the edge to the form of a sine curve of any desired wave-length; when this is reproduced it will give a pure tone of constant frequency.   Or we can draw a curve which is compounded of several simple harmonic curves, and cut this curve on the edge of the strip, and so obtain a profile representing a note of any desired musical quality.   When these artificially produced records are used for reproduction, the corresponding sounds are emitted by the loudspeaker.   Obviously this technique has great possibilities, since it is open to the creative

---

[1] *A Fugue in Cycles and Bels*, John Mills, Chapman and Hall, 1936, p. 141.

musician to cut records corresponding to sounds which have never yet been heard on land or sea.

It will be noticed that the width of film on which the sound is recorded for the purpose of the cinema is only 3 mm, the remaining 23 mm being required for the picture. It is remarkable that so narrow a strip should be capable of producing all the gradations of sound which are actually given in the course of a performance.

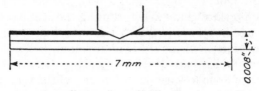

FIG. 12.7.—Recording medium of Miller Broadcasting System

When the film is developed and fixed, it is, of course, a negative, and very careful adjustment of exposure and of development is necessary to secure just the right contrast. The record must then be transferred to another film producing a positive. The sound-track is masked on unexposed film, and the picture is printed. The picture-space is then masked and the sound-track printed. This allows each to have its appropriate depth of printing. The processing of variable-density film is a very delicate matter, if the variations in density in the finished film are to correspond accurately to the variations of pressure in front of the microphone. Film development may be carried out by an automatic plant which may develop, fix, dry, and deliver as much as 75 km of film in 40 hours—a rate of about 2 km of film per hour.

The Miller Broadcasting system [1] uses a method of recording which eliminates the necessity of film-washing, developing, fixing, printing, &c. The recording medium is a tape (Fig. 12.7) in three layers. The bottom layer is transparent, and is similar to the film used in ordinary motion-picture film. The middle layer consists of a special emulsion through which the recording stylus can cut freely. Like the bottom layer, it is transparent. The top layer is very thin and densely opaque. A greater or less width of this layer is removed by the cutting stylus, according as it digs more or less deeply into the middle layer. The stylus has a wedge-shaped cutting surface with a very blunt angle (Fig. 12.8), so that

[1] *Electronics*, May 1940, p. 16.

a very small vertical amplitude of the stylus gives a large variation in the width of opaque film removed. An amplitude of a few micrometres corresponds to very loud signals. The film is ready for immediate reproduction, and can be 'monitored' 30 seconds after recording by running it straight through a reproducing unit while the record is being made.

Reproduction from Film.—To reproduce the sound from the record on the film we must re-verse the process by which the record was made. In recording we passed from the variable pressure of the sound-wave to the variable light passing through the slit and fall-ing on the film. In re-producing we must pass back again from varia-tions of light to variations of pressure in the air. Just as in recording, the process took place in two

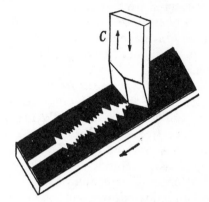

FIG. 12.8.—The Phillips–Miller method of recording

steps—(1) pressure variations to variations of electric current and (2) variations of electric current to variations of light—so in reproducing we retrace these two steps one at a time. In the first step the effective device is usually a photoelectric or 'light-sensitive' cell. This is commonly an exhausted bulb rather like an incandescent electric bulb coated with a thin film of a metal or an alloy sensitive to light. In the dark no electric current can pass through the cell, but if light falls on it a current passes which varies with the brightness of the light. All that is neces-sary, then, is to have a strong constant source of light and a slit, close to which the film record is passed at the same speed as that at which the record was made. The light coming through the slit will then vary in intensity with the variations of density of the record or the variations in width of the clear portion. In this way the variations in the light passing through the slit follow more or less exactly the variations of the light which made the record originally. The light from the slit next falls on the photo-electric cell, where it produces a variable electric current; this is then amplified and passed to the loudspeaker, where it is transformed into variations of air pressure and the original sounds are reproduced.

The reproducing unit is mounted close to the projecting unit, and the film passes first through the sound-gate and then through the projection lantern. Some slack between the two has to be allowed, since the film must pass the sound-gate with uniform speed and pass the lantern in jerks. The speed of the film is about 27 m per minute.

Another development is to take advantage of the superior quality of magnetic tape recording and to provide one or more strips of magnetic oxide on the side of the film to provide a sound track.

Frequency Requirements for Reproduction.—The ear is the only real judge of the frequency range required for reproduction. If the ear cannot detect any difference in the performance of music whether frequencies below, or above, a certain value are reproduced or not, then these frequencies are not essential for high quality reproduction. The method of test is to perform a piece of music sometimes with all frequencies below, or above, a certain value cut out and sometimes with these same frequencies left in. Observers with critical ears are then asked to express a judgement as to whether there are missing frequencies or not. If they are as often right as wrong in their judgements it is fair to assume that the frequencies are completely unimportant. This has been found to apply for piano music to frequencies below 55 Hz and above 7,000 Hz. Frequencies between 6,500 and 7,000 Hz have some importance, but only slight. Even when all frequencies below 100 and all above 5,000 Hz were cut out correct judgements reached only 80 per cent. For practical purposes it may be assumed that components, the presence or absence of which cannot be correctly judged more than 80 per cent. of the time by keen-eared listeners under good acoustic conditions, are not essential to high quality reproduction.

A study has also been made of the frequency range preference of a representative cross-section of listeners to reproduced music [1] from which it transpired that most listeners prefer either a narrow or medium frequency range to a wide one even when told that one condition is low fidelity and the other high. The result is interesting and there are three possible reasons for it. These are, accord-

[1] Olson, H., *Acoustical Engineering*, Van Nostrand.

ing to Olson: first, the average listener, after years of listening to the radio and gramophone, has become used to a more limited frequency range and feels that it is a more normal condition. Second, musical instruments are not properly designed and would be more pleasing if the higher fundamentals and harmonics were removed. Third, the distortions from true reproduction of the original sound are less objectionable with a restricted range.

The distortions and deviations from true reproduction of the original sound are as follows:

(1) Frequency discrimination;
(2) Non-linear distortion;
(3) Spatial distribution;
    (a) Relatively small source;
    (b) Separated sources in two-way loudspeaker systems;
    (c) Non-uniform directional pattern with respect to frequency;
(4) Single channel system;
(5) Phase distortion;
(6) Transient distortion;
(7) Microphone placement and balance;
(8) Acoustics of two rooms, the pick up studio and the listening room;
(9) Limited dynamic range;
(10) Difference in level of the original and reproduced sound;
(11) Noise.

In order to find out why listeners prefer a restricted frequency range Olson carried out an all-acoustic test with an acoustic filter arranged between a small ensemble and the listeners; it was then found that the wide-band condition was preferred. Thus it appears that it is the third reason mentioned above that is applicable, and tests have been made to evaluate the relative importance of the 11 possible sources of distortion. Subjective tests of non-linear distortion indicate that the amount of tolerable distortion decreases as the frequency range is increased. The tests also indicate that a very small amount of non-linear distortion can be detected when using the full frequency range.

Further tests carried out by Olson employed a stereophonic reproduction system of extremely high quality, using a very well damped room as a studio to reduce the effects of deviations (7) and (8) above. The results then indicated a preference for a full range reproduction system. The level of reproduction was arranged to be about that value found in previous tests to be preferred in smallish listening rooms. Tests have also been carried out in England involving direct comparison of reproduced and ' live ' ensembles. The results indicate that a major disadvantage of a single channel high quality system is simply that it has only one channel and the sound source as reproduced appears to be a small ' hole in the wall '.

The other deviations from true reproduction are all ones which can be removed with sufficiently high quality equipment, and so one concludes that the main reason why listeners tend to prefer a restricted frequency range in reproduced single channel music is because the spatial distribution of the source of sound is unnatural, and they become much more sensitive to any other distortions which are present. This preference disappears when a stereophonic reproduction system of high quality, carefully set up, is used.

CHAPTER 13

HALLS AND CONCERT-ROOMS

THE IMPORTANCE to the musician of the acoustic quality of the room in which he performs or listens is not disputed, although until the early years of the present century no one pretended to say why music in some rooms sounded well while in other rooms it disappointed both performers and audience. We owe the scientific investigation of the problem to the late W. C. Sabine, Professor of Physics at Harvard University, and although his work has opened up many avenues which have not yet been completely explored, the main principles are fully established, and designing for music has become a practical possibility. The fundamental fact to be grasped is that the room in which music is performed is in reality an extension of the musical instrument—voice, piano, violin, or whatever it may be—and that the tone of the instrument is modified by the characteristics of the room. Just as the tone of the reed of the oboe is modified by the associated pipe, so the tone of the oboe itself is modified by the room in which it is played, according to principles which are easy to grasp and to apply.

From the musical point of view the most important thing about a room or hall is its ' time of reverberation '—i.e., the time which sound takes to die away after the source has ceased to operate. If we sound an organ-pipe continuously in a closed room, the sound gradually builds up, the intensity getting greater and greater. If there were no escape for the sound-energy, this process would go on indefinitely. In practice it never does. After a certain interval of time the intensity and the loudness become constant, and if the source is now cut off, the sound dies away more or less rapidly. This means that somehow sound is being absorbed. While the source is maintained, sound-energy is being poured into the enclosure and the intensity is building up. But the sound-energy is also being absorbed, and absorbed at a rate which is proportional to the intensity of the sound-energy. While therefore the rate at which the energy is pumped into the enclosure is constant, the rate at which it is absorbed is at first much smaller, but gets greater as the intensity gets greater. Finally, a balance is achieved when the constant rate at which energy is put in is exactly equal to the rate at which it is absorbed, and the loudness remains constant. When the source ceases to sound, the energy input stops, but the absorption continues at a gradually diminishing rate, and finally the sound ceases to be audible. Thus the amount of absorption in the room determines three things: (1) the time taken to reach a steady loudness level, (2) the level of loudness reached, (3) the time for the sound to become inaudible. The less the room is capable of absorbing sound the longer is the time taken to reach a steady loudness, the louder is the sound-level reached, and the longer is the time during which the sound persists.

It is this last aspect of the problem which is the most important. Its application to public speech is obvious, and it was in this connection that it was first tackled. If the sound of a speaker's voice persists for three or four seconds after a syllable has been uttered, then any given syllable will overlap a whole series of succeeding syllables, and the resulting confusion makes the speaker not *inaudible*, but *unintelligible*. When complaints about not being able to hear a speaker or preacher are analysed, they generally resolve themselves into complaints that the words cannot be separated and the sense apprehended. Some intelligent anticipations of this cause of bad acoustics can be found in the literature of the subject, but they are almost completely obscured by a mass of irrelevance and ignorance. Magical remedies for bad acoustics abound. Wires stretched across the roof, specified proportions between length, breadth, and height, and many other ways have been tried and have failed. It has been said that the exact reproduction of the dimensions of an acoustically good building has been known to give an acoustically bad building. If this is true, there need be no mystery about it. We shall see that an exact reproduction *in different materials* may give an entirely different result. Sabine picked at once upon the most significant factor. He called the time for a sound to fall from average intensity to inaudibility the ' time of reverberation ', and proceeded to experiment on an acoustically bad University lecture-room, where the time of reverberation was obviously too long and all speech blurred in consequence. It may be noted in passing that a speaker may do much even in unfavourable surroundings if he speaks very deliberately and *not too loudly*.

For obvious reasons, the louder he speaks the longer the time during which his syllables persist and the greater the confusion.

In his early experiments Sabine used an organ-pipe as his source of sound and a stop-watch to time the reverberation after the source ceased. This apparatus subsequently became more elaborate, and is even more elaborate still today, but the principle remains unaltered. If the reverberation is too long, how shall we reduce it? Observant readers may have noticed the difference in reverberation in their own homes at spring-cleaning times as compared with ordinary times. The absence of carpets and curtains notably prolongs reverberation of footsteps and of the voice. Absorption then is associated with soft, porous materials; and the reason for this association is quite simple. Energy can never be destroyed, it can only be transformed into other forms. Sound is absorbed when the energy associated with the sound-waves is transformed into heat. When the compression of a sound-wave approaches a carpet, air is driven in to the pores. When the rarefaction follows, air rushes out of the pores. The friction of the air in the pores as it surges out and in generates heat energy, as motion against friction always does, and this energy, now appearing as heat, is drawn from the sound-waves, which therefore diminish in intensity, and hence in loudness. It takes a large amount of sound-energy to yield quite a small amount of heat, and the gramophone cannot be regarded as an important auxiliary method of heating a room! It has been estimated that the sound-energy generated by the shouting of a Cup-Tie crowd throughout a match would yield just about enough heat to warm one cup of tea.

Experimenting on the Harvard Lecture Room, about which he had been consulted, Sabine measured the time of reverberation with the room empty and then with various numbers of cushions brought in from a neighbouring room. He found that the time of reverberation depended on the *area* of cushion surface, and that the absorption of the empty room was equivalent to a certain area of cushion. The graph of the reciprocal of the time of reverberation against the area of cushion was thus a straight line (Fig. 13.1), and it was easy to read off from this graph the area of cushion required to give any desired time of reverberation. Furthermore, if we add to the actual area of cushion in the room the area of cushion which is equivalent to the walls, floor, ceiling, &c., of the empty room, and call this total A, then we find the very simple relationship

$$AT = k$$

where T is the time of reverberation and k is a constant for the particular room tested.

In a sense the reverberation problem is now solved for this room. But the solution is in many ways very unsatisfactory.

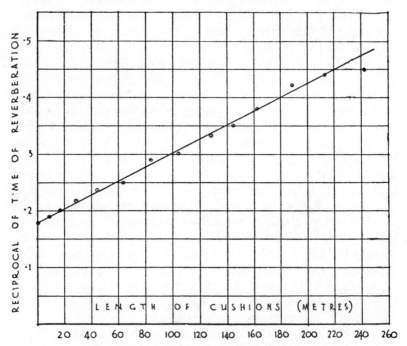

FIG. 13.1.—Straight-line relation between absorption and reciprocal of time of reverberation from W. C. Sabine's original measurements

(a) We have used a particular organ pipe and particular cushions, and the results are of very little use to other experimenters for whom these are not available.

(b) The solution is in terms of cushions, and the primary purpose of a cushion is that it should be sat upon: presumably this will alter its efficiency as an absorber.

(c) The solution takes no account of the audience, although we know this to be an important factor.

(*d*) It only helps us to correct this particular room—we should have to make the same kind of experiments in any other room which had to be corrected, because it would have a different value for *k*.

(*e*) Most important of all, it helps us with the correction of an existing room, but gives no help at all in the vital matter of designing a room which does not require the expensive and unsatisfactory process of correction.

(*f*) It offers no guidance as to the most desirable value of T.

Let us consider these points in turn.

(*a*) The organ pipe used by Sabine was rated, and found to give an intensity about one million times the minimum audible. As the time measured was that required for the sound to become inaudible, we can make the time of reverberation perfectly definite by saying that it is the time required for the sound in the room to fall to one-millionth of its initial intensity.

(*b*) We want some standard of absorption in terms of which the cushions can be measured. By a stroke of genius Sabine chose the simple standard of a square foot of open window.[1] All the sound which falls on the space of open window passes out of the room, and therefore, so far as the room is concerned, it is completely absorbed. We can find the time of reverberation in the room with a certain area of cushion surface, and then, having removed the cushions, find what area of open window is necessary to give the same time of reverberation. We thus find the area of open window which is equivalent to 1 square metre of cushion surface, and this—which is, of course, always a fraction—is called the coefficient of absorption. The total absorption of the cushions in a room is obtained in ' open-window units ' (O.W.U.) by multiplying the coefficient of absorption by the area. Obviously this process can be extended to curtains, carpets, &c. The actual processes now in use for measuring coefficients of absorption are not so simple as this, but are the same in principle.

(*c*) The process is, of course, easily extended to the audience. We only have to find how many people are equivalent to how many square metres of open window, and that enables us to calculate the absorption per person in O.W.U.

[1] Nowadays, of course, we measure the areas in square metres.

(*d*) Experiments in rooms of different volumes showed that the constant k was proportional to the volume, so that $\dfrac{k}{V}$ is a constant for all rooms and, when V is measured in cubic metres, is found to be 0·16 so that

$$\frac{k}{V} = \frac{AT}{V} = 0·16$$

or

$$T = \frac{0·16V}{A}$$

In this formula, which is the master formula for the design of rooms for music

> T is the time of reverberation in seconds,
> V is the volume of the room in cubic metres,
> A is the total absorption in O.W.U.

(*e*) But this formula can not only be applied to the correction of rooms: it can be applied to design. An architect's plans and specifications enable us to calculate V, and, from the area and coefficient of absorption of each material used, to calculate the total absorption A. We can therefore calculate the time of reverberation for the completed room before the first sod has been cut, and can make our adjustment on the drawings and specifications, instead of on the completed building.

(*f*) There are two ways in which we may attempt to determine the preferred value of the time of reverberation. One is to take a smallish room with hard walls and bare of furnishings, so that its time of reverberation is long, and, by introducing absorbents, vary the time until musical taste pronounces the room 'just right' for musical performance. This was done for a number of rooms in the case of solo pianoforte music, and it was found that musical taste was remarkably sensitive and accurate. In every case when the adjustment was pronounced correct, the time of reverberation was close to one second. This method would be very difficult to apply in the case of large halls. The alternative is to make a review of existing buildings, measure or calculate their times of reverberation, and compare the values with the judgements of musical taste on the buildings. The results obtained in this way will be considered later.

Coefficients of Absorption.—The accompanying table gives a selec-

PLATE XV

FIG. 12.9.—Domestic tape recorder.

FIG. 12.10.—Professional tape recorder.

PLATE XVI

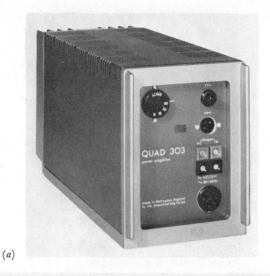

FIG. 12.11.—Typical 'Hi Fi' apparatus

FIG. 12.12.—Transcription turntable with 'High' Quality arm and pickup

tion from the almost infinitely long list of coefficients of absorption measured by a large number of different methods for all ordinary building materials, and for a great many proprietary materials devised for use as sound absorbents. The different methods of measurement lead to discrepancies in the values obtained, and these discrepancies are being further explored, but the existing figures are sufficient for all ordinary purposes.

These figures are mostly based on measurements made at the National Physical Laboratory, and are given for three different frequencies. Calculations are generally based on the values at about 500 Hz. It will be seen that ordinary wall and ceiling surfaces have very low coefficients of absorption, but that lime plaster has a coefficient about twice as great as that of hard plaster. The fact that hard plasters have so largely superseded the older lime plaster accounts to some extent for the reverberation problem in modern buildings. Floor-coverings have an absorption some ten to twenty times as great as that of plaster. Acoustic plaster, which is a special preparation and rather more expensive than the ordinary plaster, is about six to eight times as efficient as lime plaster, while the special acoustic tiles may have an absorption as great as 0·8—i.e., 80 per cent. Acoustic felts are very efficient absorbers, but harbour dust and moth and are inflammable. Fibre boards are easy to apply and fairly efficient. Asbestos sprayed on to a roof and given a white finish makes a highly absorbent ceiling. The figures for the audience per person indicate how important this factor is. One person = 0·44 square metres of open window = 0·56 square metres of best acoustic tiles = 1·7 square metres of acoustic plaster = 13·9 square metres of lime plaster = 27·9 square metres of hard plaster!

Sabine, in his original experiments, used an organ pipe of the gemshorn stop of frequency 512. This frequency is the one in terms of which the time of reverberation is usually measured or calculated. But for the performance of music the time of reverberation at other frequencies is important. Thus, if the time of reverberation is long for low-pitched tones and short for high-pitched tones, then the quality of all instruments will tend to be more mellow, and even dull. If it is short for low-pitched tones and long for high-pitched tones, then the quality tends to become more brilliant, and possibly thin. The balance of instruments in an orchestra may also be disturbed, absorption in the bass favouring violins and oboes, while absorption in the treble favours 'cellos and bassoons.

The table already given shows how the coefficients of absorp-

Absorption Coefficients of Various Materials [1]

Material.	Frequencies in Hertz		
	250.	500.	1,000–2,000.
Ordinary wall and ceiling surfaces:			
Lime plaster	0·02–0·03	0·03–0·04	0·03
Hard plaster	0·01–0·02	0·01–0·02	0·02–0·03
Unpainted brick	0·03	0·03	0·05
Wood-panelling, 3-ply . . .	0·01–0·02	0·01–0·02	0·01–0·02
Curtains:			
Cretonne	—	0·15	—
Medium weight	—	0·2–0·4	—
Heavy, in folds	—	0·5–1·0	—
Floor coverings:			
Wood block in mastic . . .	0·03	0·06	0·10
Cork carpet, ¼ in. thick . .	0·03	0·07	0·20
Porous rubber sheet, ⅛ in. thick .	0·05	0·05	0·20
Axminster carpet, ¼ in. thick .	0·05	0·10	0·35
,, ,, on ¼-in. felt underlay.	0·05	0·40	0·65
,, ,, on ¼-in. rubber ,,	0·05	0·20	0·45
Turkey carpet, ½ in. thick . .	0·10	0·25	0·30
,, ,, on ½-in. felt underlay .	0·30	0·50	0·65
Special absorbents:			
Acoustic plasters (½ to 1 in. thick) on stone	0·15	0·25	0·30
Fibre boards, plain, ½ in. thick, on battens	0·30–0·40	0·30–0·35	0·25–0·35
Medium efficiency tiles, on battens .	0·40	0·40	0·50
High-efficiency tiles, with perforated surfaces, on battens	0·50	0·80	0·85
Acoustic felts, 1 in. thick, perforated covers on hard surface . . .	0·30	0·70	0·80
Acoustic felts, ½ in. thick, on battens .	0·25	0·45	0·70
Wood wool-cement board, 1 in. thick, on battens	0·30	0·60	0·70
Sprayed asbestos, 1 in. thick . .	0·50–0·60	0·65–0·75	0·60–0·75
Slag wool or glass silk about 2 in. thick, on battens	0·70	0·85	0·90
Cabot quilt, 3-ply, two layers . .	0·40	0·70	0·70
Individual objects in open-window units (ft.):			
Audience per person	4·3	4·7	5·0
Chairs, bent ash	0·16	0·17	0·21
Cushions, hair, 2¾ sq. ft. under canvas and plush	1·1	1·8	1·5

tion vary with frequency. As a rule they increase with increasing frequency, but if the coefficient is plotted against the frequency, almost every kind of curve can be obtained by selection from the large number of materials, many of them patent materials, which

[1] See *Sound Absorbing Materials*, E. J. Evans and E. N. Bazley. Stationery Office Code 48–143, 1960.

are now available. Some of these curves are shown in Figs. 13.2 and 13.3. Some of the patent materials are said to be 'straight-line absorbents', which means that they absorb equally at all frequencies. It will be noticed that the effect of an audience is to give strong absorption in the upper register, thus reducing brilliance of tone.

Almost all absorption of sound takes place at the ceiling, floor, and walls, but not all of it is due to porosity. Sometimes a wall or ceiling may be set in vibration by the sound-waves,

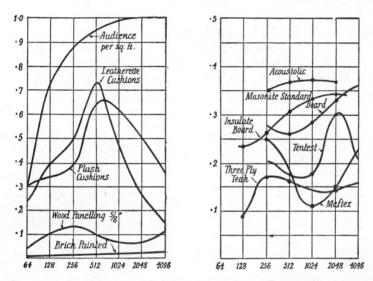

FIGS. 13.2, 13.3.—Variation of absorption coefficients with pitch

and may transmit them, thus acting effectively as an absorbent. This action is more effective for low-pitched sounds. When an organ is being played in a church, much of the bass is transmitted by the windows and can be heard outside, while the treble is hardly, if at all, audible. In addition to the absorption which takes place at the boundaries of the room, some takes place in the air itself. This absorption is almost negligible for low-pitched sounds, but increases as the pitch is raised. It depends to some extent on the moisture content of the air.

Calculation of Time of Reverberation.—As an example of the method of calculating the time of reverberation for a building for which the necessary information is available, we may take

the old Gewandhaus at Leipzig as calculated by Bagenal.[1]
This building was modified in 1842, and has always been famous
for its acoustics. On making measurements on the drawings,
we find that the area of plaster is 242 square metres. It is lime
plaster on wood lath, oil painted, and for this particular construction the tables give a coefficient of 0·02. The total absorption
of this plaster is thus 242 × 0·02 = 4·8 Open-Window Units.
The floor is of polished board, and has an area of 242 square
metres. The coefficient of absorption is 0·06, and therefore the
total absorption is 242 × 0·06 = 14·5 O.W.U. This may have
been increased a little owing to resonance and diminished a little
owing to shading by the audience, and in the final table this is
allowed for. The wood walls are boarded and painted, and have
an area of 244 square metres, and a coefficient of 0·06, giving a
total absorption of 14·7 O.W.U. Proceeding in this way, we
tabulate the permanent absorption as follows:

Old Gewandhaus.

Volume (including galleries), 2,132 cubic metres.   Seating, 570.
Volume per seat (audience), 3·74 cubic metres.

Absorbent.	Remarks.	Area (sq. m.) or number.	Coeffi-cient.	Absorp-tion O.W.U.	Adjustment.	Net ab-sorption O.W.U.
Main ceiling	Lime plaster on wood lath oil painted.	242	0·02	4·8	—	4·8
Gallery ceilings	Ditto, unpainted.	113	0·03	3·4	—	3·4
Plaster on brick	—	141	0·02	2·8	—	2·8
Glass	Negligible.	—	—	—	—	—
Wood walls	Boarded and painted.	244	0·06	14·7	Add 10% for resonance.	16·1
Doors	—	16·7	0·06	1·0	—	1·0
Floor	Polished boards.	242	0·06	14·5	Plus and minus 10%.[1]	14·5
Seats on floor	Large chairs, cush-ion seats and cane backs.	380	0·09 per chair	35·3	—	35·3
Seats on galleries	Wood and cane benches.	190	0·009 per chair	1·8	—	1·8

Total permanent absorption: 79·7

[1] The whole floor undoubtedly acted as a resonator, but, on the other hand, was shaded by audience.

We can now calculate the time of reverberation in the empty
hall. It is given by

$$T = \frac{0·16 \times 2,132}{79·7} = 4·3 \text{ seconds.}$$

[1] Bagenal and Wood, *Planning for Good Acoustics*, p. 102.

When an orchestra is rehearsing we may have fifty performers, and as a rule their chairs give very little absorption, so that the net result is to add 0·44 units of absorption per person, or 50 × 0·44 = 21·8 units in all. Adding this to the permanent absorption and recalculating, we have:

$$T = \frac{0·16 \times 2,132}{101·5} = 3·4 \text{ seconds.}$$

For the full audience on the floor (380 persons) we add 0·44 units per person, but subtract 0·09 unit per person, since the empty seats account for 0·09 units, and when the seats are occupied the absorption of the person is substituted for that of the seat.   In this way the following table is arrived at:

Absorbent.	Remarks.	Number.	Coefficient.	Absorption O.W.U.	Adjustment.	Net absorption O.W.U.
Audience on floor	On chairs as above.	380	0·44 less 0·09 per person = 0·35	133	Total audience, 570	214·9
Audience in galleries	On benches as above.	190	0·44 less 0·009 per person = 0·43	81·9		
One-third audience	—	190	Average number of units = $\frac{214·9}{3}$			71·6
Orchestra	Chairs negligible.	50	0·44	22	—	22

Reverberation T = { Full audience (570) . . . 1·1 seconds.
One-third audience (190) . . 2 „
Rehearsal (50) . . . 3·4 „
Empty . . . . 4·3 „ }

In the above case the calculation was not made in advance of construction.   In fact, it was made from the dimensions and drawings of the old hall after it had been modified.   Obviously, however, the method is applicable in advance of construction, and the following table shows the analysis for the White Rock Pavilion, Hastings, which was specially designed by Bagenal for musical requirements.

Designing for Musical Requirements.[1]—Experience shows that for any given requirement there will be a ' preferred time of reverberation '.   If the time of reverberation is too short, the room is said to be ' dead ', and the conditions are acceptable neither to audience nor to performers.   If the time of reverberation is too long, all effects are blurred, rapid musical passages become indistinct, and the conditions are generally recognized to be bad.

[1] See *Sound Insulation and Room Acoustics*, P. V. Brüel, p. 237.

Volume of White Rock Pavilion, 7,929 cubic metres. Seating
1,400. Volume per seat, 5·66 cubic metres.

Absorbent.	Remarks.	Area (sq. m.) or number.	Coeffi- cient.	Units of absorp- tion O.W.U.	Adjustment.	Net ab- sorption O.W.U.
Plaster, hard	Keene's or fibrous.	2,090	0·02	41·8	—	41·8
Wood platform, floor, and stag- ing	Oak.	109	0·06	6·5	Less 10% for shading by players.	5·8
Wood panelling round orchestra	Oak ½-in., 5-ply panels, 2-in. air space.	49	0·1	4·9	—	4·9
Wood doors	Oak, 2-in.	72	0·06	4·3	—	4·3
Glass laylight and windows	—	21	0·027	0·6	—	0·6
Carpet area on ground floor and promenades	Five-frame Wilton on thick under- mat.	892	0·25	223	Less 10% for shading by seats, etc.	200·6
Curtains	Thin.	21	0·15	3·1	—	3·1
Vents	—	100	0·046	4·6	—	4·6
Upholstered seat, arms not up- holstered	Goat hair.	1,400	0·16 per seat.	221·1	—	221·1
Settees, large, up- holstered	Seating each five people.	15	1·85 each.	27·9	—	27·9

Total permanent absorption   514·7

Audience, full	Take coefficient as 0·44 − 0·16 = 0·28	1,400	0·28	392	—	392
Audience, one- third	Take coefficient as 0·44 − 0·16 = 0·28	466	0·28	130·5	—	130·5
Rehearsal, average orchestra	Neglect platform chairs.	40	0·44	17·6	—	17·6

Reverberation T = $\begin{cases} \text{Full audience} & . & . & . & 1·40 \text{ seconds.} \\ \text{Audience and orchestra} & . & . & 1·37 & ,, \\ \text{Rehearsal} & . & . & . & 2·38 & ,, \\ \text{Hall empty} & . & . & . & 2·46 & ,, \end{cases}$

The designing of rooms and halls for music may be said to begin
when W. C. Sabine was consulted in connection with the building
of a new Concert Hall at Boston in 1898. Realizing the importance of the time of reverberation, an investigation of existing
halls was made, and it was decided to base the design of the new
hall on the times of reverberation of the old hall and of the
Leipzig Gewandhaus. The new hall was seated for an audience
about 70 per cent. greater, but the time of reverberation was
kept at about the same figure as those of the models. It is now
known that the reverberation of the Gewandhaus is less than
that given by Sabine's calculation, but it is also known that a
larger hall requires a slightly longer reverberation to give good
tone conditions. Sabine's grasp of the essential factor was
entirely vindicated, and the acoustics of the new Boston Concert
Hall are admitted, for so large a hall, to be excellent.

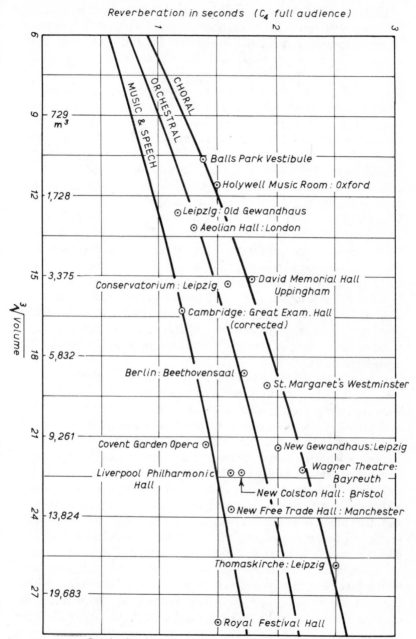

FIG. 13.4.—Graphs showing reverberation values for halls acknowledged good for musical tone

Extending this process a little farther, we can calculate the reverberation times for a number of halls which are known to be good from the point of view of music, and plot these times against the corresponding volumes or, more conveniently in practice, against the cube roots of the volume. This has been done in Fig. 13.4, in which some attempt has been made to take account of the fact that a shorter time of reverberation seems most suited to chamber music and to speech; a longer time of reverberation to orchestral music, and a still longer time of reverberation to choral music. This is, of course, what we should expect. Chamber music depends for its appeal largely on accuracy and precision—choral music depends rather on the massiveness of its effects. The Thomaskirche at Leipzig and the Wagner Theatre at Bayreuth, both recognized as excellent for choral tone, lie on the upper curve. On the other hand, some halls which are famous for good tone occur on the lower curve— e.g., the old Gewandhaus, the Royal Opera House, Covent Garden, the New Free Trade Hall, Manchester, and the Royal Festival Hall, London. It is worth noting, however, that some of these halls have a large area of resonant material, particularly wood-panelling, and the response of this material seems to compensate for a time of reverberation which would otherwise be considered short.

It is unfortunate that the London Building Act severely restricts the use of wood-panelling with air space behind. The Covent Garden Opera House was, of course, built before this Act was thought of. The importance of wood-panelling is borne out in other cases. Thus Knudsen [1] compares the Musikvereinsaal in Vienna with the Konzerthaus. The former has a very high reputation, and is said to be ideal both for performers and for listeners. The latter, while not bad, has no very high reputation. Yet the dimensions of the two buildings are not very different, and the times of reverberation are similar, that of the Musik-vereinsaal being actually shorter (1·35 seconds) than that of the Konzerthaus (1·60). The difference seems to be due to a large area of resonant materials in the case of the Musikvereinsaal, and to the fact that its surfaces are mainly smooth and hard, while those of the Konzerthaus consist of damask on the walls, thin carpet on the floor, thin cushions on the seats, all highly absorp-

[1] *Architectural Acoustics*, John Wiley and Sons (1932), p. 545.

tive for high pitch and only slightly for low pitch, thus destroying brightness of tone.

It might be well at this point to consider the requirements for the performance of music, and to deal with those from the point of view of performers and audience separately, since the requirements in the two cases are not identical. The requirements for performers are:

(1) A sense of ease and power in producing tone, and therefore an absence of fatigue. This requires reverberation, and there is some evidence that performers like more reverberation than the audience. It can be secured to some extent by keeping the reflecting surfaces near the performers hard, and therefore strongly reflecting. The way in which even the average man is stimulated to vocal exercises in the bathroom is an illustration of the importance of reverberation in sustaining tone and giving this sense of ease and power. *Punch* once carried a semi-humorous article maintaining that the Victorians had always made the mistake of furnishing their bathrooms as they ought to have furnished their drawing-rooms and vice versa. Certainly the Victorian drawing-room, with its heavy curtains and carpet and its upholstered furniture, had a very short time of reverberation, was acoustically ' dead ', and never encouraged musical effort.

(2) Good reflections from surfaces near the performers. This is secured by the suggestions already made, but it is essentially a different requirement. Performers in an orchestra or in a quartet seems to hear themselves better by reflection from a surface, and the task of an accompanist is greatly eased.

(3) Good quality of tone enhancing the specific character of the music—the richness of a bass voice, the emotional variety of the violin, the purity of the flute. This is secured by achieving the proper time of reverberation at all frequencies. It requires a balance between absorption in the low register and absorption in the high register, so that a pianist, for instance, does not have to subdue his bass or his treble, but can play with freedom. Thin curtains hung loosely absorb mainly in the upper register; thick felts, thick carpets, cushions, and upholstery absorb mainly in the middle register; resonant panelling with widely spaced studs absorbs in the lower register.

(4) Similarity between performance and rehearsal conditions. Musical performances are often spoiled by the completely different

acoustic condition of the hall on the occasion of a performance. A large audience adds greatly to the absorption, and so tends to diminish the time of reverberation. It absorbs mainly in the upper register, making the tone ' dead ' and ' dull ' and ' life-less '. It is said that on one occasion when Handel was told that he had a small audience his reply was, ' Then my music will sound better '. The effect of the audience is, of course, all the greater where the pull of box-office receipts has resulted in the hall being over-seated, and the volume per person too small. Where the volume per person is not too small, sufficient rever-

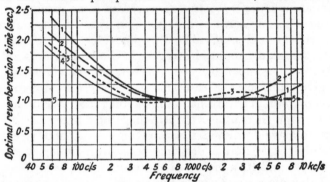

FIG. 13.5.—Curve of ideal relationship between time of reverberation and frequency. From Brüel: *Sound Insulation and Room Acoustics* (Chapman & Hall, Ltd.)

beration with a full audience can always be obtained, and the difference between rehearsal and performance conditions can be very much lessened by the use of upholstered seats, which are strongly absorbent. In this case the absorption of the audience is merely substituted for that of the seats, and the difference is not large.

Attempts have been made to derive a theoretical basis for the ideal relation between time of reverberation and frequency by starting with the variation of the sensitiveness of the ear with pitch, but at present it seems better to analyse the reverberation in rooms known to be good for tone. Fig. 13.5 shows theoretical curves relating reverberation time and frequency for rooms which are to have good tone according to various authorities; the large discrepancies at either end are immediately obvious and illustrate the difficulties of subjective assessments in this field. The reader should consult Brüel's book for further details. Absorption in the high-frequency region is nearly always great, and it is difficult to secure sufficient reverberation for the ' tops ' which give brightness to the tone. The careful polishing of all woodwork at the Leipzig Gewandhaus provides a useful hint, and the effect of polishing a sample of cork slab is found to be a marked drop in its absorption at high frequencies. According to Bagenal,[1] 'Only in the Scala at Milan, the Philharmonie and Singakademie in Berlin, and in the smaller studios of the N.B.C. in America is there recorded the desirable rise in the upper register, and it is very slight '.

(5) Absence of distracting noise. Protection against extraneous noise is an important aspect of designing for musical requirements. The most effective measure which the architect can adopt is the selection of a quiet site. In many cases this will be a counsel of perfection, but it ought to be borne in mind. Next in importance is the choice of the quiet side of a site, if such exists, and in any case selecting this side for windows. If the site is a noisy one, windows cannot serve the double purpose of protecting against noise and affording ventilation. Double-glazed windows will admit light and, if suitably designed, keep out most of the noise, but only when tightly shut. The building must have its independent ventilating system. Distracting noise from outside is mainly ' air borne '—i.e., the sounds are carried by waves in the atmosphere. But an even more troublesome kind of sound to get rid of is what is ' structure borne '. Even sounds produced in the air may generate sound-waves in the structure, running through walls, floors, and ceilings, and communicating their vibrations to the air at all points. But these structure-borne sounds are worst when they are produced by impact on some part of the structure or directly communicated to it. Footfalls on a floor above, the working of a lift, the opening and closing of lift gates, the ventilating fan—all these give rise to air-borne sounds against which the necessary protection is close-fitting windows and doors and special treatment of ventilating ducts. But they also generate structure-borne sounds, and here the remedy is more difficult. Broadly speaking, it lies in discontinuities. Every time sound-waves pass from one medium to another their intensity is diminished and some of the energy reflected. In this respect the older brick-and-mortar construction compares very favourably with the more modern ferro-concrete construction. This latter transmits waves from point to point with very little loss. Effective methods of prevention of structure-borne sounds are expensive and involve double walls, floating floors, and suspended ceilings, but there are many

[1] *Practical Acoustics*, Methuen (1942).

other partially effective expedients open to the architect. The great success of the Royal Festival Hall (Fig 13.7) shows that a good hall can be achieved even when the site is superficially most unsuitable with a railway running right by. The musical problem is probably most acute in the design of a set of rooms for music practice or for listening to gramophone records. It may, in fact, be safely said that most gramophone records are purchased after being heard in circumstances in which any real artistic judgement is out of the question. Double walls structurally independent if the cost is not prohibitive, stout walls instead of partitions in any case, tight-fitting doors and windows, no air communication along hot-water pipes, insulation of some hot-water pipe joints with rubber—these are among the points to which the architect must give consideration if he is to protect the music listener against the distractions of extraneous noise.

Of the conditions just enumerated, those which interest the audience directly are (3) and (5). The interest for the audience which corresponds to (1) is

(6) Adequate loudness. To some extent this depends on the reverberation. The loudness of a source is greater if the absorption is less, and this effect is further enhanced by the longer reverberation which goes with the smaller amount of absorption. Clearly, however, this is merely one factor in the preferred time of reverberation, and for any given hall is not under our control except to a small extent. It is important, therefore, that the source of sound should have adequate power. This is to some extent in the control of the performer, and the ear is fortunately very tolerant; but some solo instruments can never be made adequate in a large hall, and large choirs added to large orchestras may in some halls be absolutely deafening. There is an optimum number of performers for any given size of hall. To calculate this number we find the number of units of absorption necessary to give the preferred time of reverberation in the hall, and then use the empirical fact that on the average one orchestral instrument can feed about 18·6 units of absorption. Thus, taking the middle curve of the graph in Fig. 13.4 as giving the preferred time of reverberation, we can calculate the optimum number of performers as follows:

For a volume of 765 cubic metres the preferred time of reverberation is 1·0 seconds,

$$\therefore 1\cdot0 = \frac{0\cdot16 \times 765}{A}$$

where A is the required absorption.

$$\therefore A = 122 \text{ units,}$$

$$\therefore \text{ Optimum number of performers} = \frac{122}{18 \cdot 6} = 7 \text{ (approx.).}$$

Proceeding in this way we derive the following table:

Volume of room in cubic metres.	Optimum period in seconds.	Required absorption in O.W.U.	Optimum number of players.
765	1·0	122	7
1,812	1·25	232	13
3,540	1·4	405	22
6,116	1·6	612	33
9,713	1·8	863	47
14,500	2·0	1,160	63
20,650	2·5	1,322	71

(7) Absence of echoes. Echoes may on occasion be trouble-some to the performers, but they are more likely to be trouble-some to the audience. An echo is, of course, the reflection of waves from a boundary, and in order that it may be noticeable it must be (a) sufficiently intense, (b) separated from the direct wave by a time-interval of not less than 1/15th second. This latter condition is due to the fact that if the reflected wave arrives within less than 1/15th second after the direct one, it is not separately perceived by the ear, but acts merely as prolong-ing and reinforcing the original sound. Now, a *time* difference of 1/15th second corresponds to a *path* difference of about 22 metres. This means that the reflecting surface must be at least 11 metres away. Echoes are therefore not usually troublesome except in large buildings. The reflection from the back wall is the most likely source of echo, and this is a reason for using absorbent on the back wall. The reflecting surfaces near the performers give useful reflections, and cannot give echoes, since the per-formers are so close to the surfaces that the direct and reflected waves follow one another closely. Distributing absorption over the back of the hall and leaving the surfaces near the platform bare has the additional advantage that it tends to make the platform end more ' live ' than the back, and, as has been said, there is evidence that the performers prefer more reverberation than the audience. The strength of the reflected waves depends, of course, not merely on the coefficient of absorption of the

surface, but also on its form.   Sound-waves get less intense as they travel, for they are always diverging and spreading their energy over greater and greater areas.   If they are reflected from a plane surface they continue in the new direction with the same divergence as before.   If reflected from a convex surface, their divergence is increased.   Therefore convex surfaces almost never produce a strong echo.   On the other hand, reflection from a

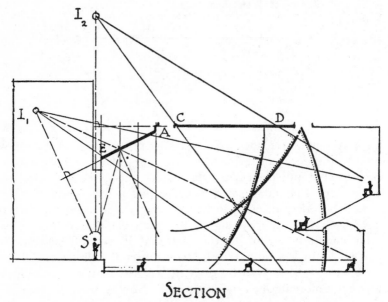

SECTION

FIG. 13.6.—Acoustic design of overhead splay. I_2 is image of speaker formed by reflection from ceiling CD; I_1 by reflection from splay EA

concave surface always diminishes the divergence of the waves, and may even make them converging, focusing them over a comparatively small area.   In this case a strong echo may appear, and concave surfaces, arising from a circular plan or dome or barrel-vault ceiling, are always sources of danger. They are admissible in some circumstances if carefully designed, but the importance of careful design cannot be too strongly emphasized.   No amount of absorbing material will completely eliminate the echo from a badly placed concave surface.   This leads us naturally to another requirement of the audience:

(8) Uniform distribution of loudness.   This depends almost entirely on the intelligent use of reflecting surfaces.   Useful

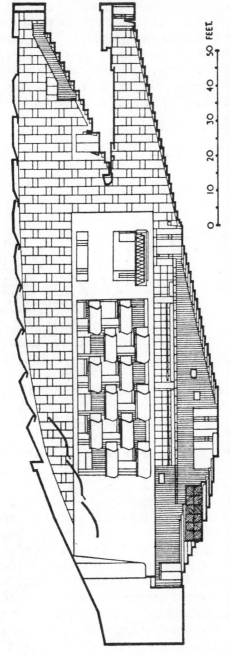

FIG. 13.7.—Royal Festival Hall, London. From Parkin and Humphreys: *Acoustics, Noise and Buildings* (Faber & Faber, Ltd.)

reflecting surfaces are those which send the sound by a direct path to the audience. The simplest way to plot the reflected wave is to take account of the fact that just as an object in front of a mirror gives an image behind the mirror from which the light-rays appear to come, so a reflected wave always appears to come from a point on the normal to the surface through the source as far behind the reflecting surface as the source is in

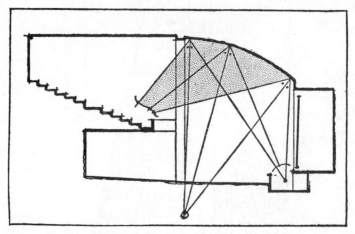

FIG. 13.8.—Undesirable concentration of sound by a concave proscenium

front. Fig. 13.6 shows, in addition to the direct wave, the reflected wave from a splay suitably designed, and the reflected wave from the ceiling, both reinforcing the direct wave where it is most needed—in the gallery and under the gallery. The effect of bad designing is shown in Fig. 13.8, where the concave proscenium focuses the sound over the front few seats of the gallery. The worst design for a domed ceiling is that in which the centre of the curved surface lies on or near floor-level. The effect of this is shown in Fig. 13.9 (a). The reflected wave is focused, and if the height is great enough, the echo would be intense. On the other hand, if the radius of the surface is twice the height of the building, the sound is well spread over the floor (Fig. 13.9 (b)). The Albert Hall suffers both from its dome ceiling and from its circular plan, and Figs. 13.10, and 13.11, show this focusing effect with long path differences. Recently the hall has been greatly improved by increasing the absorption on the curved balcony fronts and reducing the effect of the ceiling by

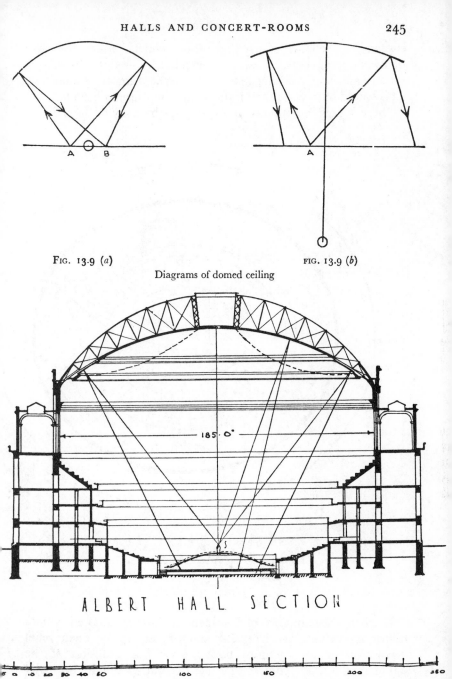

Fig. 13.9 (*a*) Fig. 13.9 (*b*)

Diagrams of domed ceiling

185·0°

ALBERT HALL SECTION

Fig. 13.10.—Cross section, Albert Hall, London

suspending 'clouds' fairly low down which reduce the time delay of the echoes (see p. 241) but which do not reduce the volume greatly and hence do not make the reverberation time too short. This would have happened if the ceiling had simply been lowered as there is now much more absorption in the body of the hall.

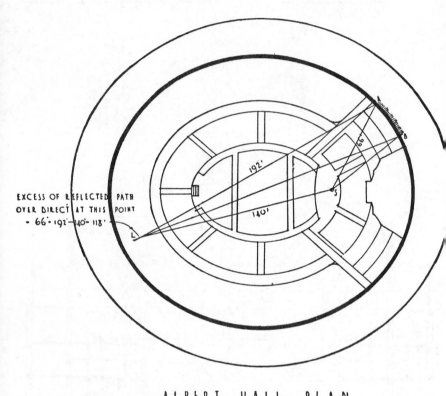

EXCESS OF REFLECTED PATH
OVER DIRECT AT THIS POINT
= 66'= 192'-140'= 118'

ALBERT HALL PLAN

FIG. 13.11.—Plan, Albert Hall, London

Uniform distribution of loudness is further assisted where reflecting surfaces are irregular and broken up by deep relief decoration or by columns, boxes, &c. The diffusion of the sound in this way requires that the decoration should be really deep—i.e., of the order of half a metre at least.

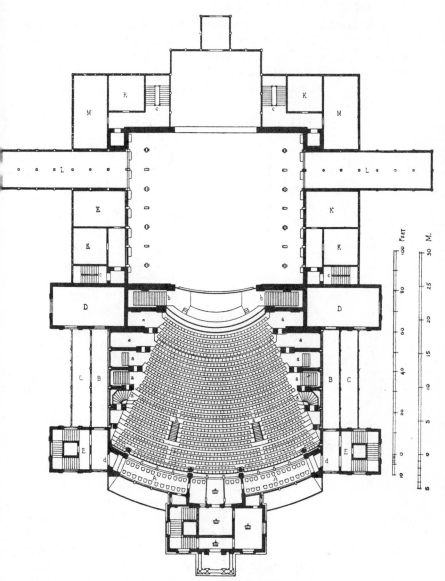

FIG. 13.12.—Bayreuth, plan of Wagner Theatre

Designing for Special Requirements

Opera-Houses.—In opera-houses the time of reverberation tends to be short. This is not good for choral tone, but it does make for intelligibility. The time of reverberation when empty is not more than 1·6 seconds in the Royal Opera Houses at Berlin, Vienna, and Budapest, and only slightly longer at Paris, Leipzig, and Dresden. With an audience the time of reverberation does not exceed 1·3 to 1·4 seconds. The traditional form is that of the Italian Opera House of the eighteenth century. It has a flat ceiling and a splay reflector over the proscenium. The auditorium wall consists largely of draped loges or boxes which are highly absorbing. A good deal of wood is used in the construction. In the Royal Opera House at Covent Garden, which preserves the old tradition, there are 9,200 feet of wood panelling. The Wagner Opera House at Bayreuth represents a complete break with the old tradition. Its time of reverberation is 7·4 seconds when empty and 2·25 when full. It has a fan shape, and the audience is carried on a steep ramp. For Wagner's music the conditions are said to be excellent, and music-lovers and critics are agreed in giving it the highest praise. But it is also admitted that for opera of the more intimate kind it is not so suitable. The plan is shown in Fig. 13.12. This theatre shows a break with tradition in another way. It has a very deep orchestra pit extending well back under the stage as shown in Fig. 13.13. The pit takes 130 performers, and with brass and drums at the bottom gives good blending. The orchestra no longer interposes itself between the audience and the chorus, par-

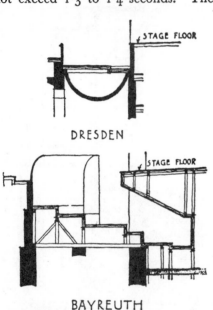

DRESDEN

BAYREUTH

FIG. 13.13.—Two types of Orchestra Pit

tially submerging the chorus, but instead combines with it
and supports it. The fact that the Wagner Opera Theatre
enhances Wagner's music, but is considered unsuitable for
Mozart or Verdi, emphasizes the relationship of music to the
building in which it is produced. It is further illustrated by
Bach's music. His choral music can be completely spoiled by
performance in small and overcrowded halls. Rapid passages
are clear and articulation is good, but the general effect is unsatisfactory. On the other hand, what may be called 'church
conditions' undoubtedly enhance choral tone, and lend beauty

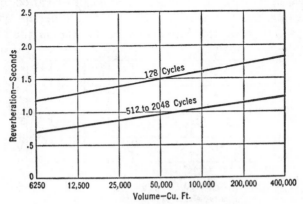

[Reprinted by permission from *Architectural Acoustics* by
Knudsen, published by John Wiley & Sons, Inc.]

FIG. 13.14.—Curves of reverberation times recommended for broadcasting and
sound-recording studios

of tone to violins in slow sustained passages. If, however, the
time of reverberation is very long, as in a cathedral or in King's
College Chapel at Cambridge, rapid parts of some of Bach's
fugues are almost unplayable. Most, if not all, of his large
works were written for production in the Thomaskirche at Leipzig.
Here he had 'church conditions', the time of reverberation being
about 2·5 seconds. If his works are performed with less reverberation, the tone suffers; if with more reverberation, the
articulation of both words and music suffers.

Broadcasting Studios.—These present a special problem, as there
are two superposed reverberations—that in the studio and that
in the room of the listener. In addition, of course, the music
is picked up by a single microphone instead of by two ears, and
the effect of listening to a radio set with two ears is quite

different from the effect of listening to the original music with two ears. It is now pretty generally agreed that the reverberation time for broadcasting studios should be less than that for concert-room conditions—probably about two-thirds. Fig. 13.14 shows the values for reverberation times at frequencies of 128 and 512–2,048 for rooms of different sizes, as recommended by Knudsen. Fig. 13.15 shows the reverberation curve for the large studio of the National Broadcasting Company of America. It is natural that the multiple use of broadcasting studios should suggest the desirability of having the amount of absorbent adjustable so that the reverberation curve can be varied and the reverberation time adjusted to speech or to different types of music. Attempts at this have been made, and in the case of

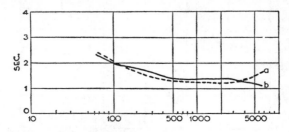

FIG. 13.15.—Reverberation curve for large studio of N.B.C. of America. *a*, curve aimed at; *b*, curve achieved

the Main Studio of the Hungarian Radio Company at Budapest are said to have achieved considerable success. According to the published accounts of this studio, the entire walls and ceiling can be covered with deeply folded hangings, any or all of which can be rolled up on rollers. By exposing part or all of these hangings the reverberation time can be varied from 0·5 second to 4·0 seconds. This provides for speech, chamber music, orchestra or band, oratorio or cathedral organ.

Brüel[1] mentions a studio design providing variable acoustics by having the walls constructed of large vertical, equilateral prisms, whose sides are lined with different absorbing materials. By turning the prisms the absorption desired can be obtained. Another adjustable studio is described in Knudsen and Harris,[2] who show a photograph of a studio belonging to

[1] Brüel, P. V., *Sound Insulation and Room Acoustics*, Chapman and Hall.
[2] Knudsen and Harris, *Acoustical Designing in Architecture* Wiley (1950), p. 395.

Station KSL at Salt Lake City, Utah, which has rotating wall panels.

The problem can also be attacked by introducing reverberation artificially between the receiving microphone and the loudspeaker. There are many ways of doing this, and experimental work is still proceeding, with encouraging results. There always remains the difficulty that as a rule the performers prefer a longer time of reverberation than the listeners, while these methods can only add more reverberation to what already exists.

In the case of the large concert-room the alteration of the time of reverberation for direct listening presents much greater difficulties, and very little progress has so far been made in this direction.

Probably there is no direction in which co-operation between the musician and the scientist is likely to be more fruitful than that of designing auditoria for music. If the musician will only learn a little of the theory and relate this to his immensely valuable experience as a performer or listener, his criticisms and suggestions would be of great importance to those who are wrestling with the actual problems of design.

BIBLIOGRAPHY

Backus, J., *The Acoustical Foundations of Music*. Wiley.
Bagenal, H., *Practical Acoustics*. Methuen.
Bagenal and A. Wood, *Planning for Good Acoustics*. Methuen.
Brüel, P. V., *Sound Insulation and Room Acoustics*. Chapman and Hall.
Carse, A., *Musical Wind Instruments*. Macmillan.
Culver, C. A. *Musical Acoustics*. Blakiston.
Helmholtz, H., *Sensations of Tone*. Longmans.
Jeans, Sir James, *Science and Music*. Cambridge University Press.
Leipp, E., *Acoustique et Musique*. Masson.
Miller, D. C., *Sound Waves, their Shape and Speed*. Macmillan Co.
Miller, D. C., *The Science of Musical Sounds*. Macmillan Co.
Parkin, P. H., and Humphreys, H. L., *Acoustics, Noise, and Buildings*.
    Faber.
Richardson, E. G., *Acoustics of Orchestral Instruments*. Edward Arnold.
Taylor, C. A., *The Physics of Musical Sounds*. EUP.
Winckel, F., *Music, Sound and Sensation*. Dover.
Seashore, C., *Psychology of Music*. McGraw Hill Pub. Coy.
Wood, A., *Sound Waves and their Uses*. Blackie.

INDEX OF SUBJECTS

INDEX OF NAMES

Science Paperbacks 100

Alexander Wood's
The Physics of Music
(Seventh Edition)
Revised by J. M. Bowsher

The new edition of this classic work which deals with that
fascinating interface between physics and music in an essentially
non-mathematical way retains the character of its predecessors
and provides all students of music with a lucid and entertaining
introduction to the subject.

Starting with a gentle exposition on the nature of wave motion,
the author then discusses sound, its generation by vibrating
systems, its reception by the ear and finally its comprehension
by the brain. This survey enables the student to understand the
next section on musical instruments (including the human voice)
where characteristic sounds obtained from the major instruments
are explained in simple terms.

The next section deals with the theories of consonance and
dissonance and hence the development of the four main systems of
temperament used in musical scales. Modern notation has been
introduced here, together with the correction of minor
inconsistencies of earlier editions.

The reproduction of music through the recording or broadcasting
chain is discussed and there is a useful section on the acoustic
behaviour of concert halls and how to attempt a correct tonal
balance, which will enable the musician to discuss intelligently
and make recommendations on the acoustic behaviour of his
surroundings.

Also from Chapman and Hall

Acoustics and Vibration Progress
Volume 1
Edited by R. W. B. Stephens and H. G. Leventhall
1974, 243 pages
Volume 2
1976, approx. 230 pages

Ultrasonic Communication by Animals
G. D. Sales and J. D. Pye
1974, 281 pages

A complete list of Science Paperbacks is available from the publisher

Science Paperbacks are published by
Chapman and Hall
11 New Fetter Lane London EC4P 4EE